Klondike '70

Klondike '70

The Alaskan Oil Boom

Daniel Jack Chasan

PRAEGER PUBLISHERS

New York · Washington · London

PRAEGER PUBLISHERS
111 Fourth Avenue, New York, N.Y. 10003, U.S.A.
5, Cromwell Place, London S.W.7, England

Published in the United States of America in 1971
by Praeger Publishers, Inc.

Library of Congress Catalog Card Number: 74–129138

Portions of this book were first published in *Esquire*.

Printed in the United States of America

Cover photograph by Joe Rychetnik

To Barbara

—not because it's good enough
but because I may never be able
to offer anything better

Klondike '70

1

ON MARCH 13, 1968, the Atlantic Richfield Oil Company, familiarly known as Arco, struck oil on the North Slope of Alaska, between the wild, rugged Brooks Mountains and the Arctic Ocean, in the area near Prudhoe Bay. (I've read that particular fact so often that by now it sounds to me a little like *Genesis.*) Three months later, Arco announced that it had struck oil again in the same area. After Arco's second strike, it became generally known that the North Slope contained one of the largest oil fields in the world. "No one knows how much oil there is," reported the *Wall Street Journal* on November 20, 1968, "but all indications point to an extremely large amount. Studies made this year at the request of Atlantic Richfield and Humble [Atlantic Richfield's partner in the exploratory drilling] by DeGolyer & MacNaughton, a Dallas-based firm of geologists and oil consultants, evaluated the area around the initial North Slope strikes as 'one of the largest petroleum accumulations known to the world today.' The firm said the area 'could develop into a field with recoverable reserves of some five billion to ten billion barrels of oil.'

"A field that size would be the largest in North America, probably surpassing the five-billion-barrel East Texas Field, discovered in 1930. But that might not be the end of it. Oil experts say that other sizable pools well might be found in the vicinity. 'I don't know of any major sedimentary basin with only one field,' says E. R. Scott, president of DeGolyer & MacNaughton.

"Alaska's Governor Walter J. Hickel, who has seen geologists' reports on the region, has estimated that the total North Slope contains some forty billion barrels of oil. This amount would put the total region almost in a class with Kuwait's Burgan Field, the world's largest, with estimated recoverable reserves of fifty-five billion to sixty billion barrels."

Hickel's latest estimate is 100 billion barrels. The oil companies, which doubtless have their reasons, aren't telling anyone how much oil *they* think is there, but twenty billion barrels seems dead certain, and in March, 1969, *U.S. News and World Report* said that an estimate of forty billion would be conservative. " 'It's another Kuwait,' say the optimists of the oil industry," *Business Week* reported, "and they just may be right."

The fact that oil lay beneath the North Slope was hardly news. The Eskimos, who had hunted seals and whales and walruses along the coast and caribou in the interior, had known about the oil for generations. They had seen it seeping out of the ground and into the water. They had even used a little of it to burn in their moss-wicked lamps. White men had known about the oil since at least the early nineteenth century, when Thomas Simpson, an explorer for the Hudson's Bay Company, found deposits along the Canadian part of the North Slope. Some fifty years after Simpson's discovery, a U.S. Navy explorer found oil along the upper Colville River. In 1914, according to an article by William R. Hunt in the February, 1970, issue of *Alaska* Magazine, William Vanvalin, a U.S. Bureau of Education teacher in Wainwright, found a whole lake of oil. "Having heard reports of an oil lake," Hunt wrote, "Vanvalin hitched up his reindeer and made a 550-mile trip to investigate. He could smell the substance when he got within a mile of the site, then finally it came into view. 'The sight that filled my eyes was most gratifying indeed. Two living springs of what appeared like engine oil with the black beds winding over and down the hill.' "

In 1923, President Warren G. Harding set aside a large part of the North Slope west of Prudhoe Bay, where Arco was to make its discovery forty-five years later, and designated it Naval Petroleum Reserve Number Four. American oil companies tried to beat the Navy to the punch—"I saw different oil companies stampede to stake oil land a year or two before it was declared the Navy Oil Reserve," 72-year-old Alfred Hopson, of Barrow, remembered in 1969—but they weren't quick enough. They hadn't made any improvements on the land, as is required for the establishment of claims, and hence they couldn't claim it. The Navy had just changed over from coal to oil, and it was busily stashing away oil reserves for its private use in emergencies. (Or for use by enterprising citizens. The Teapot Dome scandal, which occurred under the Harding Administration, involved the leasing of a naval reserve in Wyoming to private industry.)

The executive order by which Harding created a naval reserve on the North Slope said:

"Whereas there are large seepages of petroleum along the Arctic Coast of Alaska and conditions favorable to the occurrence of valuable petroleum fields on the Arctic Coast and,

"Whereas the present laws designed to promote development seem imperfectly applicable in the region because of its distance, difficulties and large expense of development and,

"Whereas the future supply of oil for the Navy is at all times a matter of national concern,

"Now . . . I, Warren Harding . . . do hereby set aside as a Naval Petroleum Reserve all of the public lands within the following described area. . . ."

The Navy started looking for oil in its Petroleum Reserve Number Four ("Pet Four," in the vernacular) in 1944, and eventually announced that it had found a field estimated at 100 million barrels. In view of the high costs of drilling and transportation in the Arctic, that wasn't enough oil to be worth developing, so, ostensibly for that reason, the Navy gave up and went home.

When the Navy stopped drilling at Pet Four, Alaska was still a territory, and all of it was owned by the federal government, as it had been since 1867, when Secretary of State Seward bought it from Russia. In 1959, Alaska became a state, with the right to select 103 million acres (out of the territory's 375 million) for its own. One of the state's first choices was the area around Prudhoe Bay—and clearly not because of the scenery, which is invisible during the sunless winter days and monotonously flat when you can see it. So far the state's selection of land has been only tentatively approved by the federal government, and the North Slope Eskimos— until oil exploration began, the only people who had ever used the land—have laid claim to the same area. But that hasn't stopped the state of Alaska from behaving as if it had clear title to Prudhoe Bay. In 1964, '65, and '67, the state leased a total of 900,000 acres on the North Slope to the Atlantic, Richfield, Humble, and Sinclair oil companies. In 1967, Atlantic Richfield (Atlantic and Richfield had merged by then) and Humble together spent $4.5 million drilling a test well that turned out to be dry. Then, in 1968, their Prudhoe Bay Number One rig struck oil.

Was Arco simply lucky where the Navy had been unlucky?

The recent history of the North Slope raises several potentially unpleasant questions. If there was so much oil under the Slope, why did the Navy stop drilling before it found the oil? The standard answers are that the Navy just didn't know that much oil was there; that it had been drilling only to verify the existence, not the size, of oil reserves in Pet Four; that it had already spent a lot of money and saw little reason to spend more. But there are people who suspect that the drilling stopped because the American oil industry didn't want the Navy to be the organization that discovered a really big North Slope field. There's evidence that the oil industry opposed naval drilling in Pet Four well before 1953, when Secretary of the Navy Robert B. Anderson advised that it be stopped, and that the industry had subsequently exerted some pressure to have Pet Four opened to drilling by private companies.

Even if one takes the standard explanations at face value, two questions remain. If the Navy stopped drilling because the evidence indicated that there were no large pools of oil, why did private industry spend so much money to see for it-self? And if private industry was convinced all along that there was plenty of oil on the Slope, why did it wait so long to move in?

A man who spent part of World War II in Alaska as a correspondent and who covered the story of the Navy's entry into Pet Four has told me that the people who opened Pet Four were chiefly oil-company employes who had found them-selves in uniform for the duration of the war. They were the same people who had opened the oil fields in Saudi Arabia. And they were convinced at the time that the North Slope had more oil than Saudi Arabia. After the war, they went back to their old jobs. The oil companies, therefore, had employes who thought very highly of the North Slope as early as the end of World War II.

And apparently their conviction was based on more than intuition. In March, 1970, William T. Pecora, director of the U.S. Geological Survey, told a House subcommittee:

"The Geological Survey, in cooperation with the U.S. Navy Petroleum Research group, did extensive work *in the late 1940's and early 1950's* on Naval Petroleum Reserve Number Four. . . . These geological studies . . . located not only sizable reserves of petroleum and gas for future use, but also gave a geological analysis of the three-dimensional jigsaw puzzle which is this part of Alaska below the surface. *At the same time,* our airborne magnetometer surveys and other remote sensor surveys . . . made studies to show the depth and thickness of the sedimentary formations and the depth to the basement. This led the private sector then to move eastward from NPR-4 to the stretch of ground between NPR-4 and the Arctic Wildlife Range. The seismic surveys of that area showed structures that were explored, and this is where the discovery of petroleum and natural gas has been made." (Emphasis added.)

Having heard the former correspondent's story and read Pecora's testimony, I wondered why the oil companies had waited so long, and why they went in exactly when they did. In April, 1970, I had a chance to talk briefly with Robert O. Anderson, chairman of the board of Atlantic Richfield, so I asked him about it. "Well," Anderson told me, "the oil business is primarily a business of exploration," and there are few places on earth that haven't been cased by oil-company geologists. That didn't strike me as a very satisfactory answer. I tried to press Anderson, saying that maybe oil companies did explore everywhere, but everyone had *known* there was oil on the North Slope. "Everyone has always known it all along *after* you make a strike," he said. Then he added something vague about new techniques without which drilling on the

Slope would have been impossible. When I visited the North Slope later that week, I asked about the new techniques. My question elicited only blank looks.

I didn't feel that I had learned much about the timing of oil exploration on the North Slope, but I figured I wasn't going to learn much and just let the matter drop. Then a very plausible source gave me a very plausible explanation that tied my odd bits of information and suspicion together and provided a neat, if still hypothetical, answer to my main question. The story was this: the major oil companies had indeed known, or at least believed very strongly, that a lot of oil lay beneath the North Slope. However, they also knew that there was oil in Texas, Louisiana, California, the Middle East, the Caribbean, and various other places. The companies were simply afraid that a heavy flow of Alaskan oil would glut the market and drive down prices. In one of the first big articles on the North Slope oil discovery, *Business Week* reported in February, 1969, that:

"Oil producers are sweating over the possibility that when cheap Arctic oil begins invading the market in 1971 or 1972 the present price structure—and the state production controls that help maintain it—may crack. . . . No one—certainly not the major oil companies that will determine the pace of Alaskan oil development—wants a repetition of what happened after the East Texas find of the 1930's."

So the companies did a very logical thing. They made a gentlemen's agreement that no one would go after the North Slope oil. They kept the agreement until well into the 1960's, when Atlantic Richfield broke it unilaterally and found oil. The other companies had to follow suit, and the rush was on. A very useful little story.

If we accept that story, the question still remains why Atlantic Richfield broke the agreement exactly when it did. The answer would seem to lie with Robert O. Anderson. Before 1965, Anderson was a big-time New Mexico cattle rancher, the largest private landowner in the United States, but he didn't have much of a hook into the oil industry except for owning the small Hondo Oil Company. (I've been told that Anderson still likes to refer to himself as "a cattle rancher who happens to own some oil interests.") Then, in 1963, he sold Hondo to the Atlantic Refining Company, which until then had been a very conservative and somewhat less than dynamic operation. One condition of the sale was that Anderson become a member of Atlantic's board of directors. Soon he was chairman of the board. (The story I heard—from one of Anderson's Atlantic Richfield employes—is that Anderson went to a board meeting, sat through the scheduled agenda, then stood up and said, "There's one more item of business: I'm going to be chairman of the board.") Shortly thereafter he announced that he wanted to increase Atlantic's domestic production, at least partly through exploration in Alaska.

The Richfield Oil Company had pioneered Alaska, striking oil on the Kenai Peninsula, near Anchorage, in 1957. Richfield also held promising leases on the North Slope. Atlantic and Richfield merged in 1966. The new Atlantic Richfield Company started drilling on the Slope so soon after the merger that the press release announcing the drilling came out on an old Atlantic Refining Company letterhead. When Anderson became chairman of the board of Atlantic, *Business Week* noted that he was a loner who had little use for the consortiums that are a standard institution of the oil industry.

Whether or not Anderson's role in the North Slope story is as Macchiavellian as some of the facts and stories might suggest, I don't know; in any case, I'm not sure it's more morally reprehensible to break an industrial agreement that is very much in violation of the federal antitrust laws and just incidentally soaks the American public than it is to keep such an agreement. But that's all conjecture. What's certain is that Atlantic Richfield did go in and did strike oil, and that the amount of oil was estimated very quickly to be at least five billion barrels.

Neither the oil companies nor the state of Alaska was slow to translate barrels into dollars (although a lot of people think the state wasn't quite quick enough to keep itself from getting badly taken). The state had sold its first 900,000 acres' worth of North Slope leases for a total of $12 million. On September 10, 1969, it sold leases on another 450,000 acres for a cool $900 million and a promise of a 12½-per-cent royalty on whatever the state and the oil companies decide is the gross value of all the oil that is recovered.

(In August, right before the lease sale, Representative Lloyd Meeds, of Washington, a member of the House Interior Committee, asked Alaska's Governor Keith Miller why the state had decided to sell leases to the tentatively approved land around Prudhoe Bay "in light of the written protests covering this land which have been on file by the native groups for some time." Miller replied, "The decision was made for several reasons, primary of which were the successful finds of Atlantic Richfield-Humble Oil in that area. Enthusiasm generated by these finds made it abundantly clear that needed revenues of substantial proportion would quite possibly be received by the state from a competitive bid for the oil and gas rights to such lands. The leasing of such tentatively approved land is specifically authorized by Section 6(g) of the Alaska Statehood Act. It should also be noted that more than three quarters of these lands were put up for competitive lease by the state on July 14, 1965, and went unclaimed.")

The oil companies had already started to move men and equipment into the North Slope and were doing so at such a rate that in the month of April, 1969, the airport at Fairbanks, which is the jumping-off place for the Slope, handled more freight than any other airport in the world. After the lease sale, big Hercules transport planes brought still more men and equipment onto the Slope, and the oil companies, which hoped to have North Slope oil on the market by 1972, pushed ahead.

Drilling in the Arctic isn't easy. The temperature at the North Slope on occasion drops as low as 70° below zero. It seldom gets that cold, but 40° below is no rarity, and the temperature is less of a problem than the almost constant cutting wind. The ground is frozen solid in the winter, but in the summer months it can become a vehicle-swallowing muck. No one drilled on the Slope during the summer of 1968 because the oil companies hadn't yet figured out how to work there in warm weather (the temperature soars to the 90's in summer) without having their equipment disappear into the mud. The North Slope is almost inaccessible. Supplies can be brought in by barge during a few weeks of the year, but when the Arctic Ocean is frozen, as it usually is, almost all supplies must be brought in by air. There is no railroad. (In fact, the entire state has only 500 miles of railroad, none of it north of Fairbanks.) Except for their airstrips, the drilling sites are completely isolated. The men who work there have six weeks on the job and two weeks off. During the six weeks on, they work twelve hours a day, seven days a week, but they work about half as efficiently as they would in Texas or Louisiana. Drilling costs, which run from $50,000 to $70,000 per well in Texas, run from one million to four million dollars per well at Prudhoe Bay.

But drilling in the Arctic shouldn't be considered an unmitigated hardship. Oil companies pride themselves on drilling in difficult locations, and an operation in the Arctic is a kind of intra-industry status symbol. It would also seem to be useful for advertising purposes. One of Atlantic Richfield's recent ads, a double-page spread, features a long blue photograph of an isolated drilling rig on a desolate snowy plain, captioned "High Noon in the Arco Circle." "It's no cinch drilling for oil anytime," reads the text of the ad, "and when you have to drill for oil in ground frozen 1500 feet deep where it's pitch dark for six months of the year, it's backbreaking. That's how it is for us at Atlantic Richfield, producers of Arco-brand petroleum products. We're drilling for oil on the North Slope of Alaska, 200 miles from the Arctic Circle. The place we call the Arco Circle. It means keeping lights on twenty-four hours a day to see what we're doing. At times, we need to run engines twenty-four hours a day so they don't freeze up. We need specially hardened steel drills and piping that won't snap like twigs in the extreme cold. And we need the kind of men it takes to work for twelve hours straight under these conditions. Atlantic Richfield is the company behind your local Richfield dealer. And as long as we have the men and equipment, he'll have the best gasoline and service for your car for years to come."

Getting oil out of the ground is less difficult than getting it out of the Arctic. Oil is often found in godforsaken places, but transportation is seldom a problem. The North Slope, with no railroad and the nearest ocean frozen most of the year, made transportation the oil companies' main concern. The state of Alaska wanted them to build a railroad that could haul supplies in and oil out and could just incidentally be used to open up a lot of interior Alaska for development. But the companies had been thinking about a pipeline at least since 1964, and that's how they decided to take the oil out. They would pump the oil to a port that was free of ice all year, then load it onto tankers. The closest year-round port was Valdez, southeast of

Anchorage, which had been devastated by the earthquake of 1964 but had been relocated and rebuilt. In October, 1968, Arco, Humble, and British Petroleum formed the Trans-Alaska Pipeline System (TAPS) to plan and build the pipeline to Valdez.

The pipeline, scheduled to be finished in 1972, will be forty-eight inches in diameter and 800 miles long, the largest crude-oil pipeline ever built. The oil companies plan to pump 500,000 barrels a day through it at first. By 1974 or '75, when they expect production at the North Slope to be fully developed, they plan to pump two million barrels a day. The first oil will go to the West Coast. Then, when the market in that region has become saturated, some of the oil will be sold in the Midwest and the East. The proposed methods of getting it there draw liberally on both history and science fiction. One possible route is the Northwest Passage above the Canadian mainland, with ice-breaking tankers used to batter a way through. (In 1969, Humble sent the tanker *Manhattan* through the Passage as an experiment.) Another possibility is to use submarine tankers that could go under the ice. That may be pretty far-fetched, but Arco has General Dynamics working on the problem now. It might also be possible to route tankers around Cape Horn or to Central America, pipe the oil across the isthmus, and send it in more tankers up to the East Coast. Finally—and, along with ice-breaking tankers through the Northwest Passage, this is the most likely way—the oil might travel by tanker from Valdez to someplace on Puget Sound, then be piped east from there. Or the oil might just be piped directly from the North Slope down through Canada to Chicago.

The oil companies' race to locate petroleum and get it out of the ground has been widely referred to as the great Alaskan "oil rush." The temptation to draw a parallel with the Klondike gold rush of 1897 has proved almost irresistible (as witness the title of this book, chosen before I had learned very much about the subject). In both cases, men flocked to the coldest, least accessible part of the United States to gouge out minerals worth fabulous amounts of money. But the parallel is false. The discovery and development of oil fields on the North Slope is less like the start of a nineteenth-century gold rush than like the exploitation of a wealthy new colony. No bearded sourdoughs are pulling sled-loads of provisions over the Brooks Mountains to the Slope. Instead, very well-heeled corporate prospectors have moved in with the assurance of government support to funnel the wealth back to the home country (that's not so far-fetched—even Alaskans tend to think of Alaska as separate from the United States), where it will presumably help to weight the balance of international politics on the home country's side. In the sixteenth century, Spain sent explorers to Mexico and South America to search for gold with much the same end in mind. In those days, it was generally accepted that the possession of gold was vital to a nation's power. Today, it is generally accepted that the possession of, or at least ready access to, petroleum is vital. For the big industrial nations, oil *is* power.

Most journalistic accounts of North Slope oil development have given a prominent place to its political significance. "It is a new ball game in the international politics of oil," *U.S. News and World Report* proclaimed, explaining that "The United States no longer faces the immediate prospect of having to rely on the erratic Middle East for large supplies of petroleum." *U.S. News and World Report* is only one of several publications that have quoted Walter J. Levy, often identified as "the dean of American oil consultants," to the effect that "a world power which depends on potentially reluctant or hostile countries for food and fuel that must travel over highly

vulnerable sea routes is by definition no world power." The
implication is that the United States will no longer have to
depend for its oil on "potentially reluctant or hostile countries."
That implication is false. To begin with, North Slope oil
fields won't get into high gear until 1974 or '75, and a lot can
happen in the Middle East between now and then. North
Slope oil won't, for example, make it possible for us to sell
the Israelis a fleet of B-52's before the next presidential election. In fact, as things now stand, there's little chance that
North Slope oil will ever—unless the United States economy
undergoes some pretty radical changes—make us totally independent of the Middle East. North Slope oil may *lessen* our
dependence on the Arab countries. And it may give us a reserve that we can use in certain emergencies to supply ourselves and our allies. This means that in a Middle Eastern
crisis, we could maneuver with relatively little fear of what
would happen if the Arab states suddenly stopped selling us
oil. Presumably, if any Arab state that depended on oil for its
income was more bellicose than we liked, we could tell it to
go to hell and wait out the crisis with no likelihood of oil
famine. One prominent Democratic senator told me he wel-
comed the North Slope discovery because he would be very
happy to have the United States depend a little less on the
"A-rabs."

Needless to say, the oil companies haven't sunk millions of
dollars into the North Slope for purely patriotic reasons. The
oil represents not only power but profit. The April, 1969,
issue of *Fortune* quoted an oil executive to the effect that
North Slope oil should yield a profit of 90 cents a barrel in
Los Angeles and 80 cents a barrel in Chicago. Using the
lower, Chicago, figure, this estimate means that by 1975, the
oil companies expect to be making a *profit* of $1.6 million

a day from the North Slope. Assuming that there are only twenty billion barrels under the Slope, the oil there represents a potential profit of at least $16 billion. The companies are understandably eager to start getting their hands on that money. Money in the hand is worth a lot more to them than money in the ground.

Probably, says *Fortune,* "Only Standard Oil Co. (New Jersey) can afford to be relaxed about Alaska. Through its Humble subsidiary, Jersey Standard has a 50-per-cent interest in the Prudhoe Bay leases that Atlantic Richfield will develop, but Jersey Standard also has plenty of oil elsewhere. It would prefer to see Alaskan production develop slowly, for rapid development might cut into markets and prices for its oil from other areas and enable competitors to fatten on Alaska profits." It's probably no coincidence that the two oil companies with large international operations—Humble and British Petroleum—are the ones interested in sending tankers through the Northwest Passage. A tanker is handy because you can send it anywhere to pick up oil and to deliver it, while a pipeline can travel only from one fixed point to another. (The *New York Times* reported in November, 1969, that, although the choice between tankers and a pipeline would be made on the basis of economics, it would not be made in an atmosphere of social-scientific detachment. "The corporate in-fighting is as rough as any seen in executive suites in recent years," the *Times* said. "Internal corporate empires and personal futures are riding on the ultimate decision.")

Atlantic Richfield, which has its main interest in American oil and American markets, has favored a pipeline from Puget Sound to Chicago and New York. The company is already building a $100 million refinery at Bellingham, Washington, ninety miles north of Seattle, to handle the first installments of North Slope oil, and it would probably like to have all the oil flow right past its door. Also, Robert O. Anderson has long been frankly interested in drilling offshore in the Pacific Northwest (the state of Washington is deciding right now, amid considerable public controversy, whether or not to let oil companies drill in Puget Sound), and if Arco found oil there, it might be handy to have a pipeline already built and ready to move it.

A preference for pipelines isn't all that distinguishes Atlantic Richfield's corporate policy from Humble's. Unlike its sometime partner, Atlantic Richfield wants the oil to come out as soon as possible (which is another reason to favor the proven technology of pipelines over the experimental technology of ice-breaking tankers). "Atlantic Richfield is in a hurry," *Fortune* says. "The merger of Richfield's West Coast facilities and Atlantic Refining's East Coast facilities in 1966 created a fast-expanding company that is hungry for profits from Alaskan oil. And now that Atlantic Richfield has absorbed Sinclair and acquired its Midwestern gas stations, the new nationwide Atlantic Richfield marketing network has an even greater need for Alaskan oil, because Sinclair was a crude-short company. . . . If the Prudhoe reserves do amount to ten billion barrels, then Atlantic Richfield, as half owner, stands to make a profit of about $5 billion. Since a quick $5 billion is worth more than a slow $5 billion, Atlantic Richfield hopes there will be no prorationing or restriction of production in Alaska."

Prorationing, restriction—in other words, government control—is where power and profit come together. If the government didn't exert its power in the oil industry's behalf, the companies involved in the North Slope could expect very little profit. American oil can't compete commercially with oil produced abroad. Without any price fixing, oil produced in the Middle East would sell for $1.50 a barrel less on the East

Coast than oil produced in Louisiana or Texas. This wouldn't help the American oil industry, so prices are fixed, through a quota system that limits the amount of foreign oil that can be imported into the United States. The quotas may be replaced by tariffs, but some form of control will certainly remain. (The result of the controls is, of course, that Americans pay much more for oil and gasoline than they'd have to if the oil industry were a genuine free-enterprise system.) There have been suggestions that North Slope oil will be able to compete with oil from the Middle East—a statement by Michael Haider, former president of Jersey Standard, that "we'd like to put some North Slope [oil] into Western Europe if ice-breaking tankers prove feasible" drew a lot of attention—but the oil companies aren't telling anyone their own ideas on the subject, and the indications are that North Slope oil will need fixed prices in order to be sold. Haider said in November, 1969, that if quotas were eliminated, "Alaskan development would slow down, although companies which already have North Slope acreage would go ahead and drill it." Still, Atlantic Richfield executive vice-president Louis Ream, speaking in Seattle one month later to 600 Northwestern businessmen, concluded by departing from his prepared text and in effect telling his audience that if they wanted a chance to profit from North Slope oil (they did; that's why they were there), they should go out and fight to preserve the import quotas.

The fixing of oil prices is justified not only as a favor to American business. It is justified also on the basis of our old friend "national security." It *is* a favor to business, to be sure, but it also provides the "incentive" without which American oil companies wouldn't bother to find and develop the domestic oil fields that the United States must have in case foreign sources of supply are cut off.

Although North Slope oil may in fact make the United

States less dependent on the Middle East, it won't provide the kind of security that the quota system was set up to ensure. Having oil fields in Alaska isn't quite the same, strategically, as having them in the "lower 48." Dr. John Blair, chief economist of the Senate subcommittee on antitrust and monopoly, a man who has watched the oil industry closely, albeit with little affection, for many years, likes to point out that import quotas were introduced largely as a result of America's experience in World War II, when the vulnerability of seaborne shipments to submarine attack became painfully clear. The desire to avoid nautical supply routes was the reason preferential treatment was granted to oil shipped *overland* from Canada and Mexico. But, Dr. Blair says, Alaska is part of the continental United States only politically; physically, it is clearly separate. North Slope oil will travel overland only as far as Valdez; after that, it will be carried in tankers that have nothing but open water between them and the Siberian coast.

Besides, the voice of logic—and more than one Congressman —asks, if the point of seeking out oil reserves in the United States is simply to *have* the oil and be able to use it if foreign sources and supply routes are closed to us, why do we rush to get it out of the ground and to market? The oil companies reply that one shouldn't be misled by simple logic. For one thing, they say, you can't start producing oil overnight. For another, as Louis Ream said after his Seattle speech, "You can't just leave oil in the ground now and expect to get it ten years later. If they try that, they'll go to open the box and discover they've lost the key." What he was presumably referring to is the fact that once wells have been drilled, underground gas pressure begins to decrease, and it becomes more difficult —and expensive—to get the oil out. But it is also true that oil left in the ground is still there ten years later, even if you can't get some of it out. Oil that has been burned in millions of automobile engines and power plants is simply gone.

There seems little likelihood that the oil companies will be forced to sell North Slope oil on an open market or that they'll have to leave the oil in the ground until it's needed at some future date. Still, the oil companies dislike uncertainty, and they're working hard to make sure that neither of those particular nightmares comes to pass. The oil industry's basic pitch is that oil, and the resourceful men who discover, extract, and sell it, is a national treasure and should be protected. The discovery of North Slope oil has been described in predictably messianic terms. In 1968, Fred L. Hartley, president of Union Oil, said, "With ever-increasing demand, we'll thank God we've found another source of supply."

At this point, the expressions of thankfulness are probably concentrated among the executives and major stockholders of the companies that hold oil leases on the North Slope, and among those Alaskans who believe that some of the oil money will come their way. The North Slope discovery hasn't yet had any spectacular effect on the national economic picture (except that the price of Atlantic Richfield stock doubled right after the discovery was announced) or on international politics (except that Canada and Venezuela, both of which count on selling their own oil in the United States, have been very unhappy).

The discovery *has* sharpened one political conflict, the struggle of Alaska's original inhabitants, the Eskimos, Indians, and Aleuts, to gain formal title to some of the state and financial compensation for loss of the rest. And it has created another conflict, the fight of conservationists, who believe that the oil industry will destroy the ecology of arctic Alaska, to keep the pipeline from being built and, in some cases, to keep the oil from coming out. In Alaska, which has only 286,000 people but has always been too impoverished to provide even those few with much in the way of roads, hospitals, or other public facilities, the miraculous appearance of $900 million and the promise of 12½-per-cent royalties plus a severance tax, an additional charge levied on every barrel of oil that comes out, have evoked visions of tremendous change and have dominated political thinking.

Even before a drop of the oil leaves Prudhoe Bay, the North Slope discovery has turned out to be a very political event.

2

THE ONLY WAY to visit the drilling sites on the North Slope is to go as an employe or a guest of one of the oil companies working there. If you go uninvited, your plane can land at one of the two airfields, but you can't venture onto the oil-company leases.

I visited the North Slope as a guest of Atlantic Richfield. Tom Brennan, a big, round-faced Irishman with a Down-East accent who is Atlantic Richfield's public-relations man in Anchorage, picked me up at my Anchorage motel at 6:30 A.M. In the car with him was Arville Schaleben, an associate editor of the *Milwaukee Journal*, who was also going up to the Slope. We drove to a hangar at the Anchorage airport and boarded Atlantic Richfield's company jet, an eight-passenger DeHavilland 125, where we were joined by Robert Atwood, the publisher of Alaska's largest and most influential newspaper, the *Anchorage Times*.

The little jet took off promptly. As we climbed out over the water, Atwood pointed to a spot on the shoreline and said, "That's our new sewage-treatment plant. It's sure messing up a lot of beautiful ecology." Atwood made a number of other wry comments about ecology during the trip. He and Brennan agreed that although the Alaskan wilderness should be essentially preserved, it was ridiculous to consider every rock and tree precious, to be protected at the price of stopping all economic development. Both made the distinction, as oilmen and Alaskans who favor oil development often do, between "conservationists" and "preservationists." They agreed that most Alaskans were "conservationists" by their definition: they wanted to keep a lot of the state unscarred to use for hunting and fishing and camping but didn't oppose a reasonable amount of economic activity in the state's undeveloped regions. The "preservationists" they defined as people who didn't want any of the wilderness destroyed for any purpose, and these people they considered impractical and unreasonable.

A layer of clouds hid the ground soon after take-off, but halfway to Fairbanks we saw a mountain peak jutting up through the clouds to our left. It was Mount McKinley, the highest peak in North America. Schaleben said he had a friend on the mountain, a World War II aviator who had crashed to his death there and had never been brought down. Atwood said that when people died on McKinley, they usually did stay there.

We couldn't see much after McKinley, but a bit north of Fairbanks we could tell that we were over a wide, frozen river, which Brennan identified as the Yukon. Atwood pointed downstream to our left and said that some people wanted to build a dam there that would eventually back up a lake as big as Lake Erie. The preservationists as usual opposed the project, he said, on the grounds that a lake would destroy duck-feeding grounds and would drown a lot of moose. They failed, he said, to take into account the new duck-feeding grounds that would probably be created, and the fact that the lake would take thirty-five years to fill; he didn't think even moose were dumb enough to stand still until they drowned. There were very sound economic reasons to oppose the project, Atwood said—

chiefly, the fact that there was no demand for the power that would be generated at the dam—but the preservationists' arguments were just silly.

There was a fifth person in the cabin, a weathered man in his fifties wearing a gray work shirt and a red hat. Tom Brennan introduced him as Charlie Guion and told Guion our names. A little later, Guion came around and showed us a sheet of paper on which Atwood's name was written correctly and mine and Schaleben's incorrectly. He asked me about the spelling of my name and I told him how to change it. Then he asked Schaleben how to spell his name. He explained that he always found it easier to remember someone's name if he wrote it down.

It turned out that Charlie Guion was an expert in noticing and remembering. Tom Brennan told us that Guion was a "scout" for Humble, which meant that he spied on other companies' drilling operations to see how they were working and how well they were doing. Since Arco and Humble are partners on the Slope, Arco was glad to ferry him up there. (Brennan said later that the "cream of the oil industry" is filtering into Alaska, and that Guion and Arco's scout are probably the best in the business.) When Arville Schaleben suggested that scouting sounded very much like industrial espionage, Brennan replied that perhaps it was, but all the oil companies did it, and it was an accepted part of the business. Besides, he said, none of the companies did anything really dirty, like planting spies in competitors' crews; last year two men who had worked in an oil crew tried to sell information to a competing oil company, and the company simply turned them over to the F.B.I.

Charlie Guion was going to the Slope specifically to spy on a rig operated by Gulf. When Brennan asked him if a court injunction obtained by Gulf to keep scouts from flying their helicopters right over the Gulf drilling rig had cramped his style at all, Guion said he just had to look from a bit farther away.

("Gulf Oil has the big one to watch on the Slope," the *Oil and Gas Journal* reported on March 23, 1970; "Its One Colville Delta State, now below 2,200 feet, is located on a $20.7-million lease which Gulf and BP [British Petroleum] bought last September." The *Journal* then said, "After word spread that a state sale was in the works [earlier the article said there was a "prospect for another state sale on the North Slope in 1970 or 1971"], information on North Slope wells tightened sharply. Some companies are not even announcing spudding of wells, let alone current drilling depth. Prior to that, companies were giving drilling data on wildcats [exploratory wells] to the bottom of the permafrost. ["Spudding" is the beginning stage of drilling; "permafrost" is a layer of perpetually frozen earth that extends down 800 to 1,300 feet.]

"Gulf Oil, whose One Colville Delta State has open acreage on two sides of the lease, may even go to court to protect the secrecy of its drilling operation. Scouts using helicopters have been reported hovering perilously close.")

Asked if the other companies planned to fight the injunction in court, Guion said no, explaining that it was quite likely that at least one company's scout helicopter had broken FAA air-safety rules by flying too close to a rig at least once, so going into court might not be wise. Schaleben then suggested that Guion was pretty lucky; in the old days, someone would probably just have shot at him. "Oh," Guion said, "I've been shot at—but only once." A rival oil company had been drilling on a beach, and he and another scout had crawled up to the top of a sand dune two or three hundred feet away to take a look with binoculars. It was night, and there were lights on the drilling rig. "The foreman saw the lights reflected off our binoculars," Guion said. "He pulled out his revolver and emptied it over our heads, just for devilment."

About 180 miles beyond Fairbanks, we looked down through scattered clouds and saw the sharp, snow-covered folds of the Brooks Mountains extending layer after layer below us and leading finally to a flat white plain. We were over the North Slope. As our plane started down, Charlie Guion pointed out the North Slope's closest approach to hills, frost heaves, called "pingos," that sometimes reach 300 feet. Except for the pingos, it was all flat. Below us, still very small in the distance, we could see an oil rig and a cluster of buildings. Then we saw another one. And another. We landed.

The Sagwon Airport, at which we had landed, is owned by Arco, Humble, and BP. The other airport at the Slope, Deadhorse, is owned by the state of Alaska. Someone asked about commercial flights to the Slope, and Tom Brennan said that Wien Consolidated, an Alaskan airline, flew into Sagwon six days a week, but always on charter from the oil companies. Arco had the plane on Mondays, Wednesdays, and Fridays; BP had it on Tuesdays, Thursdays, and Saturdays. On Arco's days the plane went to one side of the field, and on BP's it went to the other. Brennan said the system was a carryover from the days before the September, 1969, lease sale, when everything at the North Slope was done in such secrecy that rival oil companies avoided all contact with each other.

When we left the plane, the sky was perfectly clear, and the temperature was a relatively balmy 8° with a fifteen-mile-an-hour wind. In my $129.50 down-filled parka, guaranteed to keep me warm at 70° below zero, I was perfectly comfortable; in fact, I was a bit disappointed that the day wasn't colder. I noticed after a while, though, that if I left the hood of my parka down, my ears soon began to tingle painfully. (A fellow-passenger on a previous flight between Anchorage and Fairbanks had told me that during the winter of 1968, which he spent at the North Slope, the power went off one night in a trailer where some workmen were sleeping, and the next morning all the men woke up with frostbitten faces. The man said that in 1968, people were just learning how to live at the North Slope. He described seeing a crew freshly arrived from Kansas go out to put up plywood around the rigging of an oil derrick to give the drilling crews some protection from the wind. One young man absent-mindedly put a bunch of nails in his mouth to hold, as he might have done back home in Kansas. But the temperature at the Slope was a windy 20° below, and the nails froze fast to his lips.)

A pickup truck was waiting for us at the airport, and we drove a short distance down a very good gravel road to Arco's original base camp. The base camp was a low, sprawling, cream-colored building constructed entirely of prefabricated modules that had been flown up to the North Slope on Hercules transport planes. Several pickup trucks like ours were parked by the building, all with their motors running. (Car motors are kept running all the time at the North Slope so they won't freeze.) Brennan parked our truck and we went inside into a long, low corridor with metal-encased electrical wiring running along the center of the ceiling and bare light bulbs spaced along the wires. A short, husky man in mechanic's coveralls was walking up the corridor toward us, and a group of men in flannel shirts stood in an office near the door. A notice on the wall told what to do and what not to do about various injuries due to cold. Another notice said that anyone "desiring transportation to Fairbanks" had to check with a security guard by 5 P.M. the day before he wanted to leave. Two rescue sleds were propped against a wall. A coffee urn surrounded by plates of doughnuts stood on a table near the office. (You can walk into any building at a drilling site at any time of the day or night and find an urn of hot coffee and plates of doughnuts or cookies or sometimes elaborately iced cake.)

The bedrooms appeared to be remarkably comfortable, panelled in what was either wood or a good imitation, and there were only two men to a room. The building also contained a laundry, a theatre, and a sparsely equipped medical room. Arco keeps a medical-aid man at the Slope to deal with minor injuries but flies anyone who is seriously ill or injured into Fairbanks for treatment. (For most of the men at the Slope, airplanes provide the only reliable link with the outside world. There is radio-telephone communication, but it's not for general use, and anyway, the phone line is almost always tied up. "The engineers and geologists stay on the phone for two or three hours at a time, reading figures back to Dallas," Brennan said.) All the drilling sites have similarly well-equipped living quarters.

The newest thing in North Slope housing is the prefabricated two-story structure for 215 men located near the original base camp, which Arco had brought in by barge during the summer of 1969. (Sixty or seventy thousand tons of supplies were brought in during the brief period of open water in August. The barge caravan being readied for the 1970 season will carry three or four times as much. Arco has reserved every available barge on the West Coast, and other barges will be coming from Houston and points inland. So much can be taken in by barge that many of the Hercules transports, which at first were the only reliable means of transportation, are now sitting around idle. Arco leased one and a half Hercules planes a while ago, but it has let them go.) The building, which we visited later in the day, is orange-red and looks from the outside like a small-town apartment house. It had been ready since January, but sewage and water lines still weren't hooked up. Inside, it resembled a new student center at a well-endowed college. Unlike the other prefabricated buildings at the site, it was designed with enough waste space to prevent claustrophobia. There was a dining room and a lounge and a library, all with attractive modern furniture. The trimmings and rails and bookshelves were of real wood. The sleeping quarters, with built-in desks and bunk beds, would have been comfortable by the standards of most college dormitories. There were two corridors of rooms for ordinary workers and one of rooms for supervisors. The supervisors' rooms were a little larger, and every two rooms shared a bath. Men would be able to go from the living areas to a heated garage, a generating plant, a water-treatment plant, and various other places without setting foot outside.

Arco's new building is as homey as North Slope accommodations are likely to get. The oil companies don't plan to set up permanent towns on the Slope and have no intention of permitting workers to bring their families up, apparently on the assumption that if whole families had to adjust to life in the Arctic, the work force would turn over even more rapidly than it does now. Like the Army or an athletic training camp, the North Slope is a strictly masculine preserve. In an attempt to keep the workers from thinking too much about the joys of civilization, the oil companies have barred women from the Slope. Occasionally a group of women journalists is allowed to visit, but such visits are rare. To avoid drunken brawls and to keep the men at their most productive, the oil companies also bar liquor from the Slope. As a small compensation for the lack of women and booze, and for the inhospitable climate, the companies try to give the men comfortable living quarters and plenty of good food.

Our own lunch provided an idea of the quantity. We ate in a cafeteria with travel posters on the wall and a food counter loaded with a large assortment of desserts, two soups, pork chops, salmon loaf, cabbage, carrots, sweet potatoes, salad, ice cream, and canned fruit. Fresh fruit, cookies, and the inevitable pot of coffee, along with an electric shuffleboard game, were just outside the door. The food was excellent cafeteria fare, although less spectacular than I had been led to believe. (On the flight to Anchorage, I had sat next to a young man who had worked on the North Slope the previous summer as a radio operator for BP. He had longish hair and a moustache, and he said that while he was on the Slope, "Every day, every hour, someone said something really cutting about my hair." He said there was nothing good about life on the North Slope except the pay. Then he reconsidered. "The chow is good, too," he said. "I guess it's about the best food I've ever had in my life.")

But all that was later in the day. From Arco's original base camp, we drove to Drill Site One, to watch the process of drilling a well. From the truck, we could see many miles into the distance. The North Slope looked like a frozen version of the Texas Panhandle, all flat and white, with the horizon very far off. As from the plane, the drilling rigs jutting up from the shallow snow looked small in the distance. Huge trucks and graders looked tiny as they crawled along the network of gravel roads.

It is easy, very easy, to see why oilmen and development-minded Alaskans who have seen the oil field on the North Slope think it's ridiculous to worry about the possibility that oil development might destroy the Alaskan wilderness. The land is so obtrusively vast and the oil rigs look so tiny and isolated that it's hard to imagine that the oil industry could substantially change the land, much less devastate it. (However insignificant the cluster of rigs and buildings may seem against the huge, flat plain, the extent of the development on the North Slope can be overpowering to people who have seen the Slope empty. Frederick Paul, the attorney for the Arctic Slope Native Association, remembers, "I flew over Prudhoe Bay with a bunch of Eskimos in December, 1968. Every place you looked there were lights. Their eyes bugged out. They had been used to travelling across the land for hundreds of miles without seeing a light.")

Also, it is quite obvious that Arco, at least, is not wantonly ripping up and defacing the land on which it's working. Any construction work or exploration on the North Slope entails the danger that the treads and steel scoop blades of Caterpillars, or even the tires of lesser vehicles and the boots of men, will damage the tundra, the thin layer of vegetation that covers the permafrost. If the tundra is scraped away, as happens very easily in summer when the top layer of permafrost has thawed, the dark soil beneath it is exposed. The soil absorbs the heat of the sun's rays, which otherwise fall on the tundra or are reflected by the snow, and it begins to melt away. A relatively small gash can quickly melt into a ditch many feet deep. The permafrost isn't replaced, the tundra doesn't grow back, and the ditch doesn't heal. "Any disturbance of the plant cover triggers permafrost melt and erosion," three conservationist groups argued in March, 1970, in a lawsuit against Walter Hickel. "The process is essentially irreversible and results in permanent environmental degradation." I kept a sharp eye out for great ditches melted into the permafrost, but I didn't see any; nor did I see any piles of junk or even any papers beside the road. (Conservationists have worried a lot about debris on the North Slope because in the cold Arctic air, metal cans last for decades and even organic garbage remains well preserved for years.) What the area would have looked like if the oil companies hadn't come under intense public scrutiny and public pressure I don't know, but the fact is that the roadsides are considerably cleaner than almost any I've seen elsewhere in the United States, and there is every evidence that Arco is building its roads and living quarters and setting up its drilling rigs with care.

(That care can reach the point of obsession with small but conspicuous details, as I discovered later when we drove along a causeway built out into the Arctic Ocean. The snow-covered, frozen ocean was indistinguishable from the land except that it was lower than the shore and even flatter. Several barges were frozen in the ice beside the road. Two oil rigs were visible beyond the shoreline, which curved out to our left. Brennan said that out there, right on the shore, was Arco's first successful well, Prudhoe Bay Number One, and another successful well, Beechy Point. Far out on the ice we saw a string of vehicles that belonged to a geophysical exploration crew mapping rock structures beneath the ocean with the aid of echoes from subterranean dynamite blasts. Brennan stopped the truck and got out, saying something about "plastic apple trees." Atwood knew what he was talking about, but Schaleben and I had no idea. We followed Brennan onto the Arctic Ocean. We hadn't gone far before Brennan said he had found what he was looking for—a small litter of plastic foliage and a larger litter of discarded Ektachrome film cartons. Brennan explained that British Petroleum had taken two plastic apple trees and a group of photographers out onto the ice to shoot advertising pictures. The scattered foliage was all that remained of the trees, and the discarded film cartons were all that remained of the photographers. Brennan said the mess really madé him mad. "A man can be fired for tossing a candy wrapper out of a pickup truck," he said, "and then *public relations* comes and does this." He bent down to pick up some of the film cartons, and the rest of us did the same. Even Atwood, who had made so many sarcastic comments about the people who worried about a little bit of development in a huge, flat, deserted area the size of California, got down and picked up some of the junk. It just seemed the natural thing to do. "I never thought I'd be cleaning up the Arctic Ocean," Atwood said. We didn't know what else to do with the debris, so we stuffed it into our pockets and went back to the truck.)

This is not to say that there may not have been some scenes of destruction that I missed, or that when the oil companies and geophysical exploration crews first arrived at the North Slope they didn't make some costly and destructive mistakes—Tom Brennan said that much of the early damage was done by geophysical crews that tried to stretch the safe season for exploring on the tundra, stayed out there until after the top of the ground had thawed, and consequently ripped up a lot of tundra on their way back—but I didn't *see* anything bad.

Brennan turned off onto a loop of road that rejoined the main road about a thousand feet ahead and pointed out a group of wooden stakes which looked, as he said, "like a graveyard for lemmings," but was actually an experimental structure for growing various strains of grass and ground cover. A much larger experiment, involving thirteen strains of grass from arctic regions around the world, was also in progress. The object of the grass experiments is to find a strain of grass that can be used to reseed areas in which the tundra has been ripped away and the permafrost exposed. The native tundra grasses grow very slowly, and Arco is trying to find something that will grow much faster. Five of the thirteen grass strains grown in the experiment seem to be doing well, but it is still too early for Arco to be sure that any of them will flourish on the North Slope.

Brennan also told us that the loop of road we had turned onto was itself an experiment. All other roads, buildings, and work sites on the Slope are built on five feet of gravel, which provides enough insulation to keep surface permafrost from melting. This road consisted of only two feet of gravel plus a plastic foam that Arco hoped would insulate well enough to make the other three feet of gravel unnecessary. Atwood asked if the foam was put down in a solid layer or was mixed with the gravel of the road. Brennan said that was a secret; Arco wasn't telling anyone very much about the foam because it might have commercial applications beyond the North Slope.

We drove on to Drill Site One, where we found a tall, gray oil rig with a bright orange section near the top and a blue prefabricated building looking much like a smaller version of the base camp. Across the road from the camp, piles of supplies were stacked on wooden platforms set on top of empty oil drums so that they wouldn't be covered by snow. (The North Slope gets very little snow, but what snow there is blows around a lot, and whenever it meets a solid obstacle, it piles up in great drifts.)

Brennan parked the truck near two big "Christmas trees," tall assemblages of valves and wheels perched on short, thick pipes about one hundred feet apart, that seal off and will eventually release the oil at the top of wells. The drilling rig was positioned over a hole in a line with the "trees" and about a hundred feet away from the nearer one. The rig was on rails, so that it could later be slid along to punch two more wells in the same line. Arco expects each of the wells already dug and fitted with Christmas trees to produce 10,000 barrels of oil a day. The five wells eventually to be punched on that one drilling pad may produce 50,000 barrels a day. Using *Fortune*'s estimated profit figure of 80 cents a barrel, that means a profit of $40,000 a day, or $14,600,000 a year, from that one pad.

After putting on hard hats, we climbed some metal stairs

perhaps thirty feet up the outside of the rig, then went through a door into the work area, shielded from the cold and wind by high metal plates. Inside, the five men on duty weren't drilling but were "pulling pipe," taking the drill bit and the thousands of feet of pipe above it out of the hole. The pipe came out in ninety-foot lengths, raised by a huge orange block suspended from high up in the rig. It came out covered with gray, slippery-looking drilling mud which three "roughnecks" in coveralls —roughnecks are the laborers on a drilling rig—sluiced off with hoses or, when it was caked, chipped off with shovels. The mud is actually a mixture of chemicals which is needed to lubricate the drilling area and to suspend particles of rock and silt so that they won't clog the bottom of the hole. It costs roughly $100 a barrel, and it isn't wasted.

The huge block was operated from an instrument console run by the driller, the man in charge of the rig, who, like many men who reach supervisory positions in the oil fields, was tall, short-haired, and bullishly husky. The three roughnecks, working with big wrenches and a pneumatic pulley, separated each pipe from the one below it and either swung it out of the rig through an opening in the metal panels or put it in a corner of the drilling platform, where many ninety-foot lengths stood vertically, row on row. Tom Brennan explained that one activity of scouts like Charlie Guion was to count the pipes stored on a drilling platform so that they could figure out how deep the rig was drilling.

Overhead, the lengths of pipe were being detached from the block and moved to the storage area by a derrickman who climbed about high in the rig, exposed to the wind, swinging from girder to girder with the sureness of a chimpanzee. The whole crew was working rapidly and with great intensity, two men walking a huge wrench around the joint between two pipes, one man hurrying to start the pneumatic pulley, three men laboring to tilt a ten-foot length of pipe straight upright, forming a straining tableau like the flag-raising on Iwo Jima. It was hard, dangerous work—while we watched, one man narrowly missed being hit on the head by a swaying pipe— and it would go on for four or five hours.

It took some effort for me to realize that these mud-spattered men wrestling with pipes *were* "the North Slope oil boom." All the grandiose abstractions—the ten-digit profit figures, the comparisons with Kuwait, the wishful predictions of increased national security—and even the romantic images of man battling the elements and living at the frozen edge of the earth—rested on their twelve-hour days. There on the drilling floor, their work was the only reality. The wind and cold and the flat white surface of the Arctic, less romantic than monotonous even when we could actually see it, were all shut out by the protective metal plates around the outside of the rig. The only job at hand was lifting huge lengths of muddy pipe out of a deep hole. Everyone concentrated on that. The men could have been anywhere in the world. When they were through working, they'd go back to their prefabricated living quarters, trailer-camp America run wild, to watch movies or eat doughnuts or just sleep. They really were only a few miles from the Arctic Ocean. They really were drilling toward the largest deposit of oil on the North American continent. Their work really was going to produce billions of dollars in profit, albeit for someone else. But all that was in my mind, and I had to coax it out. The straining and concentration of the men, the gray mud and the dark blotches of grease, the vibrating hum and pneumatic spurt of motors and the clang of metal on metal were all that I could see or hear.

Roughnecks can earn close to $20,000 a year at the North Slope, three times what they can expect in Texas or Louisiana, where many of them come from, and drillers can earn up to

$25,000, roughly double what they make further south. But they all earn their pay. Adding cold weather and stark isolation to the long hours, it's easy to see why, despite the high pay, the labor force at the Slope turns over rapidly.

(Eskimos apparently have a lower rate of turnover than white workers. They are more accustomed to the cold and several thousand miles closer to their homes than experienced workers from Texas and Louisiana. Also, Atwood told me, an Alaskan anthropologist has explained their persistence by the fact that work at the North Slope is very much like the work routine of traditional Eskimo culture: In traditional Eskimo society, the men went off on long hunting trips, returned to their villages with game, and then, after all the villagers had celebrated their return, went back to the hunt; now, some Eskimo men go off on six-week trips to the oil fields, return with money instead of game, and then, in two weeks, after everyone has celebrated their return, go back to the oil fields. The only Eskimo I saw at the North Slope was an elderly man sweeping the floor of a bleak corridor, but Brennan said that Arco had hired quite a few Eskimos—he didn't know how many—to work in the drilling crews, and that since Eskimos as a group have incredibly high mechanical aptitude—the Canadian government has trained Eskimos to be jet mechanics in eight hours instead of the 640 required for everyone else— they had proved to be very satisfactory workers.)

We stayed at the rig until the roughnecks fished out the gyroscope that's used to steer the drill bit deep underground. (Oil wells aren't drilled straight down. Drilling at a 45-degree angle isn't unusual. A drill bit can be maneuvered under thousands of feet of rock with considerable accuracy.) We watched them remove the bit, three wheels studded with metal teeth, the three fanning out like the jaws of a monstrous metal snake. Then we drove back toward the base camp.

In the afternoon, after our trip to the ocean and our North

Slope lunch, we visited two more wells. The first was the Sag
River well, Arco's second success at the North Slope, which
proved that the first well was no fluke and enabled Arco to
determine that it was sitting on at least five or ten billion barrels of oil. All we could see of the well was a small "Christmas tree" protected by a shed and marked with a sign, as a
historical landmark. The Sag River discovery was the one that
really started the oil boom. Brennan said he had heard that
when the Sag River drilling rig struck oil, "they had a couple
of vice presidents on the drilling floor, laughing, singing, gig-
gling, dancing around, and getting in the way." The well now
produces a little diesel fuel, which is refined in a long, green
shed-like building next door, and a little natural gas for use
at the North Slope. It is the only North Slope well whose con-
tents are being used.

From the Sag River well, we set out for Placid State Number One, a well some twenty-three miles away that Arco was
drilling for the small Placid Oil Company. On our way, we
passed the BP base camp, and there, flanking the entrance,
were the two plastic apple trees. After a while we came to a
place where four huge corrugated-metal pipes were being
placed in the ground as culverts to take the Kuparuk River
under the road after the summer thaw. Then we turned off the
good gravel road onto another, so bad in places that our heads
bounced off the roof of the truck. The scenery was still the
same: a flat, white plain stretching incredibly far to the horizon, with widely scattered oil rigs jutting into the air and
heavy vehicles crawling along distant roads.

Brennan said that the land beside the road was infested
with little white foxes that sometimes ate the insulation off
wires laid across the tundra by geophysical exploration crews.
He added that Arco had invited a couple of Eskimos over
from Barrow specifically to trap the foxes.

The Placid State rig stood at the very end of the bumpy

road. Inside, we discovered that the crew was actually drilling, and the drill bit was down some 7,900 feet. As we watched, the supervisor, Bill Congdon, a husky man with an air of absolute competence, walked casually onto the drilling floor, then suddenly grabbed a lever and shut the machinery down. He spoke to the driller, and the roughnecks started moving around. I couldn't tell what had happened but Atwood, who had been much closer, explained, "Congdon heard a noise. He said the bit had broken." And because Congdon had heard some slight irregularity in the sound amid the regular mechanical racket, the crew began pulling the bit up from 7,900 feet.

Congdon has worked on most of the important Arco wells on the North Slope, including the very first well, the unsuccessful Suzie Number One. He told us he was absolutely sure that Placid State Number One would hit oil, probably at around 11,000 feet. (The average depth of wells in the lower United States is about 8,500 feet, although some wells have been drilled nearly four times that deep. Drilling costs in the "lower 48" run roughly fifteen dollars a foot.) This well was far enough inland, he said, so that the ground was relatively dry and permafrost wasn't much of a problem. In fact, he was drilling there as if there were no permafrost at all, except that he put twenty-inch casing around his drill shaft to insulate it all the way down to below the permafrost line. Closer to the coast, he said, drilling wasn't so easy. At the Sag River well, the ground around the hole had kept sloughing back into the hole. The men had kept pouring cement down around the shaft—22,000 bags—to hold the ground back, but the cement hadn't done much good. It was later found to have diffused through the soil as far as 100 feet from the well. Finally, the only thing that had worked was simply to keep fighting down through the ground without any special techniques. A refrigerated drilling operation hadn't done any good at all on another well, and neither had one that used the insulating properties of air. The main problem wasn't really permafrost, Congdon explained, but drilling through silt and gravel, which are found in many oil fields and which always slough back into the hole. Drilling on the North Slope close to the water is difficult but not uniquely so.

We said goodbye to Congdon and then stopped for coffee outside a long, narrow room in which off-duty roughnecks were watching a full-color movie epic of the Vietnam war. Then we filed back into the pickup and drove back down the same bumpy road toward the airport. Off to our left, a helicopter was flying very low and very slowly, slowly enough to get a very good look at something. All four of us thought immediately of Charlie Guion, the scout, and wondered if the helicopter was off to take a look at Placid State Number One.

3

I MADE MY FIRST TRIP to Alaska in January, 1970. I hadn't expected to like the state, but I had expected to find a great many visible traces of the oil boom—if not oil derricks on the outskirts of town, like the ones you see in Oklahoma, at least signs of a lot of new money and a lot of honkytonks in the cities. I was wrong on both counts.

I left Seattle for Anchorage at 7:15 on a dark, rainy morning aboard an Alaska Airlines jet that the woman at the reservations counter had told me (twice) featured "golden-samovar service." I didn't take her seriously, but again I was wrong. As soon as the plane was airborne, two stewardesses, wearing red Ukranian tunics and black fur hats, came down the aisle wheeling an unmistakable golden samovar. I was glad to see it: hot tea would make the early morning a lot more bearable. But the airline apparently thought that tea wouldn't be festive enough. The dark brown liquid that the stewardess poured into my plastic glass turned out to be a sweet Black Russian. After a few incredulous sips, I decided to wait until they broke out the tea bags for break-fast. (The historical roots of the golden-samovar routine can with some imagination be traced back to Russia's ownership of Alaska. Few vestiges of the czarist period remain except for scattered Russian Orthodox churches and congregations in the southeastern part of the state, a high percentage of Russian Orthodoxy among the Aleuts, the aboriginal inhabitants of the Aleutian Islands, and occasional aged samovars that are sold in antique and curio shops for about $350. Golden-samovar service isn't Alaska Airlines' first historical extravaganza. It was preceded by a Gold Rush motif that included tassled curtains—still in place—in the airplanes and announcements written in the style of Robert Service. P.T. Barnum at 37,000 feet. I like it.)

I put the regrettable Black Russian as far away as I could and picked up the airline's magazine. On the first inside page was a management report that said, "We have recorded the best nine months in the company's history this year with a net income of . . . 70 cents per share compared with a loss of . . . $1.67 per share for the comparable period of 1968. . . . The aggressiveness of your company is also reflected in the increase in contract and charter flights. Most of the charter revenue was earned serving the needs of the oil industry in the North Slope region of Alaska. The potential effect of the oil industry on the State of Alaska is now universally known. However, it is axiomatic, as reported by the management, that as Alaska prospers so will Alaska Airlines. The revenue earned from the charter flights has this year exceeded 176 per cent of that posted during the same period last year." A little later, the magazine described the new cargo center the airline was building in Fairbanks as part of the Fairbanks airport's general expansion for handling shipments to the North Slope.

We flew through clouds a lot of the way, and I didn't see anything of Alaska until we came in for a landing at Anchorage. Then I looked down and saw thick ice floating in gray, deadly-looking water. We made our approach over the water. To the left, clearly visible at 8:30 A.M., hung an almost-full moon.

The Anchorage airport is functional enough, but it could very plausibly have "Greyhound" inscribed over the door. Inside, there is a souvenir shop that sells the same native handicrafts that are available in all Alaskan cities: carvings in walrus-tusk ivory and stone, beaded slippers, handwoven baskets, decorative caribou-hide masks with slits cut out for eyes and mouth and a ring of fur around the outside. Just before the departure of southbound planes, little knots of men can be seen talking together, like compatriots in a foreign country. Their accents are pure Texas, and when they board the planes, they take their seats in the first-class section.

One of the first things that struck me about Anchorage, as I rode downtown in the airport limousine, was the ugliness of its buildings. The one-story frame houses were ugly. The office buildings were ugly. They all smacked of prefabrication, of money saved on materials, of bare functionality as an esthetic ideal—a logical architecture for a state in which both materials and labor are very expensive, but depressing nonetheless. Even when we got downtown, there were one-story houses thrown in amid commercial buildings, and there were big gaps between buildings, robbing the place of the compression that intensifies the vitality of downtown areas in many older cities. The most elaborate structure we passed on the way in was still being built, with scaffolding all around the sides. It was a new local headquarters for the Union Oil Company. (Two days later I rode past the Union Oil building again, and the cab driver said, "I wish they'd spend some of that money on the North Slope instead. Seems like they haven't been doing much this year." I asked him if he thought the people of Alaska really had much to gain from the oil development. "Hell, yes," he said, "the state should get rich off of it.")

The two most imposing buildings in town were hotels, the Westward and the Captain Cook. (I never sampled the Captain Cook, but I did look in on the Westward. The lobby, predominately red and highlighted by imitation wood, contained a huge globe on which both Anchorage and Seattle had been rubbed clean away by a succession of pointing fingers. One of the hotel's various special rooms was called the Petroleum Club.) I stayed at a smaller, cheaper hotel that was very clean and perfectly adequate. On the desk in my room, I found a card listing the names and telephone numbers of airlines, bus lines, hospitals, restaurants, and various other data a casual traveler might need. A full third of the card was taken up by oil-industry listings. There were three geophysical companies, sixteen oil producers, and nine drilling companies.

Even before the North Slope strike, Anchorage had been something of an oil boom town and Alaska something of an oil-oriented state. Oil was discovered on the Kenai Peninsula, just across Cook Inlet from Anchorage, in 1957, and Anchorage owes much of its recent growth and prosperity to that discovery. "Since the discovery of oil in the Kenai Peninsula in 1957," *Business Week* reported in February, 1969, "employment in the oil industry has taken up much of the slack left by a tapering off of defense-related spending. [The military moved a great many men into Alaska during World War II, and the various military bases quickly became the backbone of the Alaskan economy. The bases are still there, and so are a lot of soldiers, but the military establishment in Alaska is no longer the gold mine it was during and right after the war, and it will probably become even less so as soon as a military withdrawal can be accomplished without plunging Alaska into a depression.] Oil-industry payrolls, which grew from $25 million in 1966 to an estimated $50 million in 1969, already are a significant addition to the Alaskan economy. Lease payments, filing fees, drilling permits, and similar expenditures (adding up to $42 million in 1967) have proved so profitable that, according to W. W. Hopkins, of the Western Oil and Gas Association, they make up about 40 per cent of the state's general fund budget."

When I left the hotel, the impression I got on foot was the same as from the limousine: the gaps between the buildings, the dismal architecture. There were few people in the streets and almost no traffic. The weather was unseasonally warm: it was almost zero. (Alaskans may be used to the cold, but they don't just ignore it. They talk about it all the time.) Most of the stores were one-story buildings. The façades of many of the taller buildings sported colored plastic paneling. I didn't have to walk far to find the bars and pawnshops. Most of the former have pool tables at the back and knots of regular customers perched on the stools. Most of the latter are well supplied with guns. There are a lot of lady taxi drivers, and more Southern accents than I've heard anyplace else north of Virginia. (That's true all over Alaska. A journalist who had spent a while in Alaska told me disgustedly a month before I left that "the state is populated almost exclusively by ex-Southerners and retired military men." As a result, he believed, the Alaskan political mentality was hopelessly right-wing. Actually, Alaskans don't generally elect right-wingers to office. There's a large enough rightist lunatic fringe, however, so that anyone who publicly takes a strong liberal stand can expect to get harassing and threatening phone calls. Arthur Hippler, president of the state affiliate of the American Civil

Liberties Union, says that every time he appears on the radio or makes a public statement, the calls come in like clockwork.)

Walter Hickel once spoke, when he was still just a real-estate entrepreneur, of "building a Fifth Avenue on the tundra." This statement has been quoted many times, and I expected to find Anchorage a city that halfway fulfilled his dream. Instead, I found a drab, flaccid place that could have been any small city anywhere in the United States. Almost, anyway. When the sky cleared, magnificent mountains rose sharply behind the city. There were a lot of Indian faces in the streets. There was a wine shop that offered vintage Chateau Latour and Dom Perignon champagne. And on the front lawn of the little one-story town hall, draped in strings of colored Christmas-tree lights, was a heavy section of forty-eight-inch steel pipe, a sample of the pipe for the proposed TAPS pipeline to Valdez.

Actually, a lot is happening in Anchorage. It is very much *the* city in Alaska. It is the commercial capital. It contains nearly half of Alaska's population now, some 130,000 people, and it is growing more rapidly than anyplace else in the state. The city's population is expected to double by 1974. The oil companies are putting up buildings and moving in employes, and as a result, housing, which has always been scarce and expensive, has become almost impossible to find. Hotel rooms are also in short supply. A Seattle transportation expert, Fred Tolan, predicted in late 1969 that if North Slope oil development goes according to plan, Anchorage "will require 2,800 more housing units per year, a 15-per-cent increase in the capacity of the Alaska Railroad, more docks, more truck terminals, more airports, more warehouses, and more business stores and buildings. On top of that there will be vast increases in schools, hospitals, water systems, light and power stations, sewer facilities, and everything else."

Still, you can't tell just by walking around the town that it's growing by leaps and bounds, or that it's full of people who plan to profit mightily from the biggest oil strike on the North American continent.

Fairbanks, where I went next, wasn't exactly Gay Nineties either, but I found several obvious, rough-edged signs of the boom. There was construction at and around the airport, and the Coop Drugstore—"Alaska's largest"—was selling picture postcards of North Slope oil wells. In a bar called Tommy's Elbow Room, where the bartender smashed each beer bottle as he disposed of it, I overheard a long, detailed discussion about which construction companies were going to build which sections of the pipeline. On the plane back from Fairbanks, I sat next to a man who told me that "like a lot of people in this state, I've hitched my wagon to the oil industry."

The *Anchorage Daily News* reported in July, 1969, that, despite the oil boom, "Outwardly [Fairbanks] hasn't changed much. It still has the same old air of utter impermanence— 'like a second-class beach resort,' one colleague put it. Radio station KJNP, 'the gospel station at the top of the nation,' gives the fundamentalist version of how it is, while the fly-blown joints downtown dispense the hardest core pornography in the state. . . . This is Fairbanks, perennial boom and bust town, booming again under the impact of all those billions of barrels of oil from the North Slope. . . .

"The oil activity here revolves around Fairbanks International Airport. . . . All this activity is certainly making itself felt, but just what impact it is having on the economy is difficult to estimate accurately. 'I don't tell much difference between now and any other year,' one cabbie said. . . . 'All those guys from the Slope make a lot of money, but they don't spend it here. They just stop over at the airport, buy a fistful of tickets and keep on going.' But there's no doubt that the oil activity has turned a perennial housing shortage into a famine. Many rents have increased sharply and some have doubled. . . ."

As I walked through the streets, though, even the *Daily News* account seemed exaggerated. The *News* wrote, "Downtown the streets are crowded—with oil workers just back from the Slope or going there, with Fairbanks oldtimers, with G.I.'s and air men from the bases and with a few hippies. . . ." In January the streets weren't crowded with anyone. Winter in Fairbanks isn't an ideal time for strolling, and people don't spend much time outdoors. Many of the pedestrians I did see were natives dressed in store-bought parkas, corduroy on the outside with fur-trimmed hoods and bands of embroidery around the bottom. Neither natives nor whites often bothered to zip up their parkas or jackets in the near-zero temperature as they moved between stores or walked from buildings to their cars or pick-up trucks.

Second Avenue in Fairbanks is both the main drag and the honkytonk street, and it is often described as a wild place. I found it lined with low, decrepit office buildings, a seedy movie theater, the Nordale Hotel, the ubiquitous souvenir shops, a furrier's shop, the Coop Drugstore, a most un-Oriental-looking Chinese restaurant, the telephone and telegraph office, and a string of bars. The neon signs of the bars are the bright lights of Fairbanks. They don't differ appreciably from the bar signs that line at least one depressing street in almost every American city and town I've ever seen. Men on vacation from the oil fields do frequent the bars, and by the time the last pedestrian totters across the sidewalk it's early morning. But, like the vaunted bustle of downtown Anchorage, the boisterous atmosphere of Fairbanks' Second Avenue is a gross exaggeration.

The man-made drabness of Anchorage and Fairbanks is reasonably typical of Alaska, a state not noted for its inspiring architecture or its urban vitality. But there is more to Alaska than the cheesy houses and the dismal streets. For one thing, there is real hospitality. Whether or not that's a symptom of the frontier I don't know, but the fact is that people invite you quickly and readily to their homes. There is also a kind of frontier feeling that seems to reflect three inescapable facts: you're always aware of the vast, wild landscape outside the cities and towns (most Alaskans hunt and fish, and a great many own their own planes); you're always aware of the presence of aboriginal cultures (even if the main evidence is the artifacts in souvenir shops and the Eskimo or Indian features of most of the poorer people you see in the streets); and you're always aware of the isolation (Alaskans speak of trips to the lower 48 states as "going outside").

Now oil is in the air, too, almost as much as the wild landscape and the isolation and the aboriginal cultures. How to spend the $900 million gained from the September, 1969, lease sale and how to maximize oil revenue in the future are the big issues in Alaska. When Alaska became a state, in 1959, it had little hope of economic success. It has depended heavily on federal military bases and federal handouts, and it still hasn't been able to make ends meet. The $900 million represented six times the state's total budget for fiscal 1969. It was obviously a fantastic windfall, but no one knew exactly what to do with it. To get some ideas, the state hired the Brookings Institution, which ran a series of seminars in Alaska, and the Stanford Research Institute, which issued a thick report. The question of what to do with the money has been discussed in the legislature and in the newspapers. A surprising number of Alaskans have opinions on the proper level for oil severance taxes and the comparative virtues of competitive and noncompetitive lease bidding. Liberals all seem to have read Robert Engler's meticulous and scathing book *The Politics of Oil.* If an oil expert visits Alaska to make a couple of speeches, his words are front-page news.

It's obvious, as the Stanford Research Institute said, that "The discovery and development of substantial oil resources on the North Slope of Alaska's Brooks Range will have vast economic and social impact on the state." However, judging by the rest of what the Institute said, North Slope oil isn't likely to turn Alaska into another Kuwait: "Far from transforming Alaska into an embarrassingly rich state," the people from Stanford predicted, "the North Slope oil discoveries will only help to place the state on a more comparable basis with its sister states with regard to a sound financial position and adequate public services. The high cost differentials and other disadvantages of a relatively remote and sparsely populated land will persist and could even be accentuated unless the bonus money and other state revenues are used wisely and in the public interest."

In fact, the Institute said, things may well get worse before they get any better: "It is anticipated that costly short- and long-range demands will be made on state and local governments to provide service to the growing population in areas affected by the extraction and shipping of the oil. Indirect but important results of these added private and public expenditures that can be expected are temporary shortages of housing, labor, and other resources and some accompanying upward pressure on prices."

"A large backlog of social and developmental needs already exists in Alaska," the Institute went on; "past state revenues have been insufficient to attack these needs at a desirable pace; and, except for interest on the oil bonus money, future state revenues are not expected to increase substantially until at least fiscal year 1973, when sizable royalties from North Slope oil are first anticipated. Furthermore, one result of added government capital outlays is to increase operating expenditures, and this result must be added to future revenue requirements.

"In most populated areas that have either exhibited recent rapid growth or are experiencing unprecedented growth resulting from the impact of the oil boom, the urban services needed to support resident and working populations—water, sewer, solid-waste disposal, energy, streets, public health and safety, housing, recreation and cultural facilities, schools, and so forth —currently vary from adequate to oftentimes severely inadequate. Substandard services usually reflect an insufficient tax base to pay for the facilities and personnel required. Rapid growth will further aggravate the present situation and sorely overtax facilities and services in most of the urbanizing areas. In some instances, these shortages may create discomfort and even financial hardship on present and future residents and in other cases, public health and safety may be impaired at least until additional facilities and trained personnel can be provided."

Certainly, in terms of money in the bank, the state is and will be rich beyond its wildest dreams. But how much difference that will make—to what extent dollars in the bank can or will be translated into jobs or economic development, for instance—is open to serious question. The oil industry doesn't employ many people; even the rosiest estimates don't anticipate that the industry will have more than 11,500 employes in Alaska by 1985, and many of these people will probably be imported. Whether jobs will become increasingly available in other industries is questionable. Operating costs are very high in Alaska, and the distance to any conceivable market is

large. Alaska has always depended chiefly on its natural resources—salmon, lumber, gold, and copper. (For years, the
salmon-canning industry literally ran the state. It controlled
at least five seats in the eight-man territorial senate, and its
lobbyist, W. J. Arnold, sat in the senate chamber and told the
senators how to vote.) There's very little chance that a big
hunk of extra capital will change that. Certainly the state can
sink some of its money into public works. In a huge state with
only about 5,000 miles of roads, road-building alone could
take up a lot of the available cash and could provide a lot of
short-run jobs in the bargain. But North Slope oil money
won't even enable the state to build roads to its heart's content.
The Stanford Research Institute observed, "Even allowing for
increased state revenues deriving from the North Slope oil
boom, it appears that construction of all the added mileage
being considered for the next twenty years under 100-per-cent
state financing would place a serious strain on the state's resources."

Nevertheless, the fact remains that North Slope oil has already provided the state with a huge quantity of money that it
wouldn't otherwise have had, and future oil revenues, whatever they are, will also represent found money. Not surprisingly, a great many Alaskans, private citizens as well as
politicians, have come to identify the fortunes of the state with
those of the oil industry.

In Anchorage, the process of identification had a head start.
People thought that there was oil on the Kenai Peninsula, just
across Cook Inlet, well before World War II. The only trouble
was that the oil was inconveniently located in a national wildlife refuge, the Kenai Moose Range. In the period after the
war, which might otherwise have been a good time to move
into the Kenai, the oil industry found itself in some disrepute,
which might have made it politically risky for the government

to grant that kind of concession to a private oil company. President Harry S. Truman's attempt to appoint an oilman, Edwin W. Pauley, as Undersecretary of the Navy had stirred up a lot of political conflict and had focused national attention on efforts by the oil industry to influence government and public opinion. The Richfield Oil Company was definitely interested in the Kenai, but it wasn't willing to push very hard.

As I've heard the story, it didn't have to. Some Anchorage business and civic leaders decided (with how much encouragement I don't know) that they wanted very much to have Richfield move into the Kenai, and they had no compunctions about approaching the federal government on their own. The Anchorage people suggested to Richfield that they would try to get permission to drill in the Kenai Moose Range and then bring the oil company in. Richfield approved, and the Anchorage people went to work.

Alaska wasn't yet a state, and some congressional backing seemed essential, so first of all, the Alaskans lined up Washington's two senators, Henry Jackson and Warren Magnuson. Then they got letters of approval from all the Alaskan sportsmen's groups. But they figured something more was necessary. Whatever the military said it needed, the military got, so the logical thing, they reasoned, was to convince the military that it needed Alaskan oil. It wasn't hard to think of arguments— without Alaskan oil, the military depended on a 2,400-mile line of supply for its numerous Alaskan bases—but arguments weren't enough; it would clearly be necessary to have a military leader as a friend at court.

As it happened, that was no great problem either. General Nathan Twining was then chairman of the Joint Chiefs of Staff. As Robert Atwood explains it, "Nate had been our general in Alaska. Everybody knew him." Members of the

Anchorage group huddled with representatives of the oil
industry to prepare a brief that they could present to Twining
in Washington and to Eisenhower's Secretary of the Interior,
Henry Seaton. The oil companies offered to provide plane
tickets for the Anchorage delegation, but the men from
Anchorage wisely refused; they realized that it would be better
politics to pay their own way.

In Washington, they found Seaton receptive, and Twining
downright helpful. First he helped the Anchorage people to
impress his subordinates with all their major points; then he
called Seaton and said the military had to have that oil. The
Anchorage delegation flew back home with its mission completed. Leases on the Kenai were subsequently granted to a
number of Anchorage's leading citizens. The Richfield Oil
Company drilled on those leases and struck oil. As a result, so
the story goes, a good many of Anchorage's leading citizens
now earn $3,000 to $5,000 a month in royalties on the oil
that is taken from the Kenai Peninsula. They are quite justified
in believing that what's good for the oil industry is good for
them.

This attitude has inevitably penetrated Alaskan politics,
which is, in the best frontier tradition, extremely personal.
Personal invective and backbiting are common. It is not only
possible but likely that anyone who is seriously interested in
politics will know his United States senators and representative, as well as a good part of the state government. The newspapers cover sessions of the legislature like school papers
describing the meetings of a club: they refer familiarly to politicians' characteristic articles of clothing and dutifully record
the legislators' ostensibly witty remarks. The everyone-knows-
everyone-else atmosphere of Alaskan politics is especially no-
ticeable in the state capital, Juneau, a seedy but picturesque
waterfront town of about 10,000 people whose only major

industry is the state government. When the legislature opens
a session, the shops put signs in windows and doors saying,
"Welcome back, legislators."

An Anchorage lawyer complained to me, "When Senator
Stevens [Ted Stevens, one of the present senators] talks about
the Trans-Alaska Pipeline, he talks about 'we.' But he's not
the worst; the state government thinks of itself as the hand-
maiden of the oil companies." The *St. Louis Post Dispatch*
reported on November 23, 1969, "At times the enthusiasm of
Alaska politicians goes to the point of embarrassing the oil
companies, which must worry about public opinion in the
lower 48." (In fact, both of Alaska's senators are closely
identified with the oil industry. Senator Stevens used to work
as a lawyer for the Mobile and Union oil companies. Uncon-
firmed but prevalent rumor has it that Senator Mike Gravel
received oil-industry help in paying off his campaign debts—
and that he voted to retain Russell Long as Senate whip and to
confirm Clement Haynsworth as a Supreme Court Justice in
order to please his benefactors. He has certainly made a lot of
money by developing real estate near the site of the Kenai oil
discoveries.)

Even if the oil companies made no effort to control Alaskan
politics, their influence would be powerful and widespread.
And, of course, the oil industry has never left politics to
chance. After the Kenai Peninsula discovery, the industry sent
five or six lobbyists to the state legislature in Juneau. After
the North Slope strike, the number rose to fifteen or sixteen.
(One prominent legislator told me that the lobbyists in
Juneau didn't seem to be the oil industry's first team: most of
them weren't very smooth. And they weren't very communica-
tive, either. "They'll buy you a drink at two in the morning,"
he said, "but they won't tell you *anything*.")

So far the lobbyists have concentrated chiefly on keeping

severance taxes down and generally looking out for the details of the oil companies' most narrowly construed interests. They did intervene in 1969 to kill in committee a bill that would have required quota hiring of minority workers, but that kind of intervention has been rare. Still, as more and more Alaskans develop a stake in oil-industry profits, the lobbyists' influence is bound to become more and more pervasive. Their job is doubtless made a lot easier by the fact that conflict of interest is a concept that hasn't taken root in Alaska. After the ice-breaking tanker *Manhattan* made its way through the Northwest Passage to Barrow, the Mobile Oil Company thoughtfully shipped the entire state legislature, lock, stock, and barrel, to Barrow for a red-carpet tour of the *Manhattan*. The oil company did its best to show the legislators a good time. When one legislator was asked afterward if he didn't think the outing might have been just a little bit compromising, he was astonished that anyone could imagine such a thing. If Mobile wanted to be nice to the Alaska legislature, what could be wrong with that? During the 1969 Senate Interior Committee hearings on Walter Hickel's nomination as Secretary of the Interior, Hickel, then Governor of Alaska, was asked repeatedly about the oil-industry connections of the state Commissioner of Natural Resources, Thomas E. Kelly. At one point, Senator Frank E. Moss of Utah said to Hickel, "You stated earlier in the questioning that you had no knowledge of whether your Commissioner of Natural Resources held any oil stock."

Hickel said, "That is right."

"Do you not think," Moss asked, "you should have made inquiry on this before you appointed him as Commissioner of Natural Resources?"

"Well," said Hickel, "I knew this man. . . . I did not think about any conflict because he was not with any major company. There is no conflict of philosophy. And so I have no further comment on it, because I feel real clean and decent about it."

"Did he not also have a relationship with the Halbuty Oil Company?" Moss asked.

"He is a stepson of Mr. Halbuty," Hickel replied.

"He is a stepson of Mr. Halbuty?" Moss repeated.

"Yes," Hickel said. "You have to get these men from somewhere."

It's hard to see how Alaska can avoid becoming an oil-dominated state. Certainly the oil companies' decisions, investments, and timing will *de facto* determine the main course of Alaska's development. Joseph Fitzgerald, who was for years chairman of the Federal Field Committee for Development Planning in Alaska, an independent United States government planning and study group, has said that he accepted a job as Atlantic Richfield's community-relations director in Anchorage because, from now on in the field of development planning, it is the oil companies who will be making the important decisions; the best way to influence the future of Alaska, he feels, is to be in a position to exert at least a small influence on oil-company policies and actions. And certainly when the oil companies set out deliberately to exert influence, they will be able to pull the most important strings in Alaskan politics. Ernest Gruening, who was Governor of Alaska from 1939 to 1953 and subsequently represented the state in the U.S. Senate, is not being overly cynical when he predicts that "oil money will dominate state legislators and elected officials much as the canned-salmon moneys did in the '30's and early '40's."

4

OIL AND THE PROSPECT of oil have made the state of Alaska suddenly covet land that for centuries no one but the natives had ever wanted. They have also made the natives even less eager than they would otherwise be to part with their land. And, wonder of wonders, they have made the stakes so high that although the natives may not be able to keep much of their land, they will be paid a tremendous sum for what they lose. Of course, the sum, though tremendous, will be much lower than market value. The natives will be treated as justly as is politically feasible.

The original inhabitants of Alaska, the Eskimos, Indians, and Aleuts (who now constitute about one fifth of Alaska's nearly 300,000 people), were never coerced or tricked into signing away their land, as the Indians of the "lower 48" were. White settlers found most of Alaska too inhospitable to bother taking. The summers are short, the winters are long and cold, the soil in the northern third of the state stays frozen all year long, much of the 32,000-mile coastline is ice-bound most of the year, the whole land is broken by rugged mountains, including 20,000-foot Mount McKinley, and distances are vast. (The state is larger than Texas and California put together, stretches 1,300 miles from southeast to northwest, and spans

three time zones.) By and large, white men stayed in settlements and cities, and the natives were free to roam uncon-
tested over almost the entire territory.

The first white settlers in Alaska were Russians who came
to trade with the natives for furs. Vitus Bering, a Dane sailing
for the Czar, discovered Alaska in 1741. He was followed by
fur traders and adventurers who ransacked the Aleutian
Islands and traded with the Indians along the southeastern
coast. During the early nineteenth century, Russians made
trips into the interior and set up trading posts, but aside from
their headquarters, first on Kodiak Island and then at Sitka,
they were transients. Traders from the British Hudson's Bay
Company penetrated the interior, too, and established a trading post at Fort Yukon. And starting in 1848 whalers fished
along the coast, but they didn't take up much room. Then the
United States, under Secretary of State William H. Seward,
bought Alaska from Russia in 1867. The Americans set up
canneries and sawmills along the southeastern coast, whalers
and fur traders roamed the Arctic coast, the gold rush of
1897 drew people into the interior, and subsequent gold
discoveries led to the founding of Fairbanks and Nome; but
by and large, the Americans, too, took little of the land. Even
today, most of the people outside south-central and southeastern Alaska are natives. Until thirty years ago, before the
war and the postwar construction of the Alcan highway had
brought thousands of additional whites to Alaska, natives
were the majority all over the state.

This is not to say that even the earliest white men brought
no changes. They had a tremendous impact on all the native
cultures. When the Russians came, the population of Alaska
was probably about 80,000. The Eskimos were concentrated
in the northern part of the state. Those along the coast depended on whales, walruses, seals, and fish for their living.

Those inland, more nomadic than the coastal people, depended on the caribou. In the Aleutian Islands lived the Aleuts, who depended entirely for food, clothing, and even parts of their shelter on the fish and animals of the sea. Most of the interior of the state, the part in which living off the land was most difficult, was inhabited by Athabascan-speaking Indians who hunted, fished, and gathered plants for sustenance. Along the southeastern coast were the highly developed salmon-fishing Indian cultures of the Tlingits and the Haidas.

The Russians first directed their attention to the Aleutians, where their weapons and their diseases managed to reduce the native population from about 20,000 to about 2,000. "The Aleuts' flourishing culture and economy did not long survive [Bering's] discovery," wrote T. P. Bank in *Scientific American* in 1958. "When the Bering expedition returned to Russia and told of the vast herds of fur animals in the North Pacific, fortune hunters started a stampede almost equal to the great Klondike gold rush that came some 150 years later. Adventurers, thieves, exiles, murderers, and princes alike set out to plunder this remote region of its treasure. A tide of greed, cruelty, and bloodshed swept over the Aleutian Islands. The Aleuts fought back, but they were overwhelmed by the superior weapons of the Russian hunters. Whole villages were wiped out; the population was decimated not only by guns but also by smallpox, measles, tuberculosis, and pneumonia."

European weapons and diseases made inroads on the mainland, too. The authoritative government study "Alaska Natives and the Land," prepared in 1968 by the Federal Field Committee for Development Planning in Alaska, reported that in the area around Cook Inlet, "The 1838 smallpox epidemic . . . decimated the Indian population and broke down group morale and action, enabling Russian exploitation to expand and [to] start the decline of native culture in the region." White traders destroyed native trade patterns. "The historic gatherings at the great trading centers of the Arctic were absolutely essential to the [Eskimo] way of life and settlement patterns that existed prior to European contact," the Field Committee said, but, according to Robert F. Spencer's standard study of the North Alaskan Eskimo, "This important trading system, which had been in existence for many years, was halted with the advent of extensive commercial whaling in the north Bering and Chukchi seas beginning in 1848. Whalers transporting goods chiefly in their own ships and distributing them directly to the Eskimo effectively curtailed the native traders, who could offer little in the way of competition. This, in turn, had a dramatic effect on the inland Eskimo. Many inlanders, no longer able to obtain trade goods upon which they were quite dependent, were forced to move to the coastal villages and learn a new way of life." The goods available from traders and whalers sucked the natives into the white man's economic system, bringing them in to live around trading posts and driving them into activities that could bring them enough money to buy the white man's goods.

"Up to about 1915," recalled Alfred Hopson, a 72-year-old Eskimo, "the whaling was the business that changed the life of the Eskimo." Hopson, the son of an Eskimo woman and a Liverpool sailor who had come to Point Barrow on board a British whaler, said, "Many whaling ships came in the summer and passed Point Barrow to go into the Canadian Arctic and whaled from Herschel Island and also Bailey Island [across the Amundsen Gulf]. . . . Some whaling companies established whaling stations at Point Hope and Point Barrow. At one time there were two companies in Barrow, and many people came from along the coast and also some from the inland. The whalers supplied food and clothing for the families year around just to have them man their boats in April and

May and also in the fall. The average estimate price of the baleen [whalebone] from one whale was $10,000. Some Eskimos had their own boats and whaled independently. Three Eskimos had three crews each and caught whales, but the traders usually got the baleen.

"After a time the whalers wanted too much for the baleen, and manufacturers offered a prize of $25,000 to anyone making a suitable substitute. So about 1910, it was produced. When the price of baleen dropped, the whaling companies gave their men an allowance, and all Eskimos went into debt. . . . The whalers did not come any more, and the Eskimos had to find a living elsewhere. The price of fur increased about this time, which again caused many families to move. They used whale boats and umiaks [boats made of sealskin], and the whole North Coast was populated in one summer, from Point Barrow to the boundary line of Canada, some even going to the Mackenzie River.

"In 1930, when I took census and travelled east, there were families camped all along the coast about ten miles from each other, many places having as many as four to five families. . . . Again the fur prices went down and the people moved back, and by 1950 only the traders, with a small number of Eskimos, settled at Barter Island. . . .

"In 1944 the Navy came with the Seabees to start oil exploration, and that was the beginning of an era whereby people began to look at jobs for a living. . . ."

During the 1920's inland Eskimos moved to the coast, where they trapped foxes and other animals whose pelts they sold to traders. With the money they bought tobacco, white flour, and manufactured goods. When they could no longer get credit at the trading post, which had fallen on hard times, they moved back inland.

Farther south in Alaska, beginning in the late nineteenth

century, Americans set up canneries and sawmills, and they, too, drew natives from their old homes and their old ways of life. All over the state, white hunters and trappers made inroads into the huge populations of game animals and fish on which all the native cultures depended.

The gold miners who flooded into interior Alaska in the '90's and the hunters who supplied them made permanent changes in the number and migration patterns of Alaskan game. "The Gold Rush period greatly affected the native subsistence pattern," says the Federal Field Committee report. "Not only were there shifts in native populations, but also the impact of feeding thousands of prospectors made wildlife resources very scarce. The harvest of wildlife resources was particularly disastrous—thousands of big game animals were required by the booming mining communities, and they were extirpated from many areas of former range. The indiscriminate use of fire also greatly affected wildlife populations. Caribou migration routes changed, and these animals completely disappeared from former ranges. . . .

"The placer mining period also disastrously affected many bottom-land habitats that the native people depended upon for food and for harvest. The result of gold-washing procedures of the time was great saltation in bottom-land flats—the habitat of waterfowl, muskrat, and beaver—causing wildlife populations to either shift or disappear."

In the places where white men chose to settle, the natives lost effective sovereignty over the land. As cities built up, natives moved into them to work. Today, many natives work in the cities for a while, then go back to their villages. Alaskan cities have large transient, basically unassimilated, populations of natives.

Most natives still live in villages, though. In 1968, the Federal Field Committee observed, "Village Alaska—the 178 predominantly native places [of 25 persons or more] scattered across the state—is the home of about 37,400 Eskimos, Indians, and Aleuts, about 70 per cent of the native population residing in the state. Village Alaska stretches from the communities of Metlakatla and Hydaburg, in the rainforests of the southeastern Panhandle, north and west 1,300 miles to Barrow and Wainwright on the tundra along the Arctic Ocean, and south and west nearly 1,600 miles to Nikolski and Atka on the foggy, lushly vegetative islands of the Aleutian Chain.

"In a number of ways these places . . . are unalike—in size, in climate, in landscape, in cultural heritage and its continuing influences, and in patterns of life and work—but in important ways they are alike. Most importantly, they are alike in that village people rely upon gathered resources of the lands and waters—not upon income from jobs—as a base for their subsistence. While not all villages or all village people depend to the same extent upon hunting, fishing, trapping, and other activities of gathering for food, reliance on gathering activities is generally characteristic of village Alaska."

On my first trip to Alaska, I flew into the Eskimo village of Anaktuvuk Pass, which lies in a flat, elevated valley among the jagged Brooks Mountains. The people of Anaktuvuk Pass are inland Eskimos, whose traditional economy was based on the caribou and who still hunt the caribou, which often migrate through their pass, for food and skins. As noted, the inland Eskimos are generally more migratory than those on the coast, and Anaktuvuk Pass has existed as a fixed settlement only since 1948. Like many small, remote places in Alaska, it has an airstrip. Wien Consolidated Airlines flies scheduled flights in twice a week. There are no arrivals and departures on the same day, though, and I had no invitations to spend the night, so I chartered a small plane in Fairbanks for my trip. We flew right in among the mountains, which were very sharp and snow-covered, and when we landed, we found a group of curious men in parkas wating for us at the runway. I identified myself to one, to whom I had been given a letter of introduction, and he led me off toward his house. A couple of the men went off on snowmobiles. There were huskie dogs in the village, standing or playing or curled up peacefully in the snow (it was 22° below zero), but it was obvious that snowmobiles were the standard means of transportation. Except for a slightly larger one-story grammar school, the village consisted entirely of flat-roofed one-room houses. One or two older houses were made of logs, but all the others were plywood. The front door to every house was protected by a low, enclosed entryway. Inside, the standard set-up was a stove with a stovepipe extending through the roof, a table, beds on one side of the room, and kids all over. A pot of coffee was kept hot on the stove all the time. In the houses I visited, I saw uncompleted caribou-hide masks, trimmed with fur. The people of Anaktuvuk Pass make these masks, which have no traditional or ritual significance whatsoever, to send to the cities, where tourists buy them for as much as thirty-five dollars apiece. Outside, among the empty metal oil drums that litter the village, caribou and other hides are hung on platforms made of poles to cure. In one house, I talked with an old Eskimo named Simon Paneak, who was a great caribou hunter in his youth. He remembered the way the mountains had been fifty years ago, and when he told me about a great migration from the mountains to the Arctic coast, he pointed to an old man hunched in a chair beside the wall and said, "His grandfather led the people up northeast near Herschel Island." The epic history of the place was that close. So was a transistor radio that picked up nothing but the conversation of the pilots of passing airplanes and was kept on apparently for that purpose. Paneak told me that the young people from the village who went away to schools operated by the Bureau of Indian Affairs often returned just because they liked village life. His own grandson, he said, went away to school for training as a mechanical engineer but married an Indian girl from New Mexico and came back to Anaktuvuk Pass. "Now he's in Fairbanks earning money, and she's living here in the village." Paneak said he hoped the village would survive. The village proper may be relatively new, but, he said, "When my mother and father were kids, a lot of people lived in these mountains." When he talked about the state's taking land that the natives in other areas had used, he said the state was "like a raven that comes down and eats your meat."

Wherever they live, the natives are far and away the worst-off people in Alaska. In many respects, they are worse off than any other group in the United States. Ramsey Clark, former Attorney General of the United States, told the Senate Interior Committee on August 6, 1969, "The present condition of the native Alaskan is intolerable. His average age at death is 34.3 years, barely half the life of all our citizens. Infant mortality is nearly two and one-half times that of white Alaskans. Ninety-five per cent of all native dwellings need replacement. A survey of twenty native villages in northwest Alaska found [that] fewer than 10 per cent of the households had water wells, the only satisfactory water source. Flush toilets were found in only two villages—a single flush toilet in each village, none in the other eighteen. [Only] one fourth of the households had privies, and only half of these were adequate from a health standpoint.

"A government report finds the native population averaging less than one half the calories medically recommended and variety in food grossly deficient.

"Joblessness in the winter averages 50 per cent of the labor force and runs 80 to 100 per cent in many native villages. The average per-capita income of all native Alaskans is less than one fourth that of white Alaskans. Prices in Alaska are the highest in the nation, and prices in native areas are the highest in Alaska. Eighty per cent of welfare in Alaska for Aid to Dependent Children goes to natives.

"Educational opportunities for native youth are terribly limited. Many villages offer nothing beyond the eighth grade. In 1960, only 8 per cent [of the natives of secondary-school age] finished high school. . . .

"Perhaps no single fact more graphically dramatizes the fate of native people in the advance of our civilization than that today, from the hundreds of millions of acres they have used and occupied of the 375,000,000 in the state, 60,000 native Alaskans now have . . . [full ownership of] 500 acres. This is roughly 400 square feet each, a room twenty feet by twenty feet. This in Alaska, where hunters must still cover many townships to feed and clothe their families."

Representative Wayne Aspinall of Colorado, chairman of the House Interior Committee, suggested to Clark that despite the grim statistics, "The treatment we have been giving to all the people of Alaska during this century has led to a better way of living . . . and increased their opportunities for longer life and more education and so forth. Is this not true?" he asked Clark. "Well, you know," Clark replied, "I think the statistics and our experience and our knowledge demonstrate beyond question that the disadvantages of the Alaskan natives are immense today, and that many years ago there were many more of them living in their native way. Who is going to judge which is the better? In their native way they were thriving. They did not have our diseases and they did not have the incursions of our hunters and our fishermen and our exploiters seeking timber and oil. . . ."

The land is very important to natives of all the Alaskan cultural groups. Howard Rock, an Eskimo who is the editor of a very literate native weekly called the *Tundra Times,* says that the Eskimos have always had a great reverence for the land. "The Eskimos have named every hill and rock. When I was a boy in Point Hope, the hunters would come in and say they had killed an animal at such and such a hill with a certain big rock at the bottom, and the other hunters would know exactly where they had killed the animal. The villages are all built in the best locations, right where the animals come through."

James Thomas, a Tlingit who for a while was director of public relations for the statewide Alaska Federation of Natives, says, "I got a great respect for the land from my mother. She was always reminding me that my grandfather and great grandfather had lived there. My family has lived and fished on the river for hundreds of years. My grandfather owned twelve slaves. When he was dying, they took the slaves upstream and killed them, then floated the bodies down the river one by one past the old chief to show him that his servants had already gone to prepare the way for him." Thomas says that a feeling of possessiveness about the land is basic to Tlingit culture. "When Tlingits pray, the opening phrase is, 'O great land-owner above.' "

Athabascan Emil Notti, a thin, quiet man who has been president of the Alaska Federation of Natives since 1967, says, "I fly into villages and sit down with the village men to talk. I tell them that the state can take their land and . . . can sell it or do whatever it wants. They get angry. They say, 'My father had a camp here and my grandfather had a camp twenty-five miles over there. I inherited the land from them, and I want to pass it down to my children.' "

Cultural values aside, the land and water and their products are vital to the economic survival of whites as well as natives in Alaska. Salmon, gold, copper, lumber, pulp, and now oil—

not agriculture, trade, or manufacturing—have sustained white society in the state. Fish and animals sustained the natives' traditional societies. Now that the natives have been sucked into the white man's economic system, they want access to minerals, too. For all this, they need the land.

The natives' first attempt to get legal title to their land was probably made in 1917, when, according to the *Tundra Times,* a claim "was filed by a trader named Newton on behalf of the Tanacross natives. [Tanacross is an Athabascan Indian village in east-central Alaska.]

"Newton reportedly had the natives mark hunting trails, fishing sites, trap lines, village sites, burial grounds, and other lands relating to historical use on a map of the area," the *Times* said in July, 1969. "Chief Andrew Isaac of Tanacross indicated that the map was registered with the Territorial Commissioner for the Upper Tanana, John Hakdukovich. This claim, Isaac said, extended from Delta Junction to the Canadian border and included all the villages in that area.

"A letter from the late Senator E. L. Bartlett to Chief Isaac, in which the Senator said that he had located an old claim, is in the possession of the Tanacross village. The Senator said that he discovered an unsigned copy of an old Tanacross claim. It is believed that this was a copy of the claim filed in 1934 by Tanacross. This claim is reported to have been dismissed by the Interior Department, which said the claim was too large.

"*Tundra Times* research revealed the existence of another early claim, made in 1946. Chief Isaac said that a man who identified himself as Judge Goldstein, from the Bureau of Indian Affairs' Native Legal Service, came to Tanacross and had the villagers mark maps. All lands claimed through historic use and occupancy by the Tanacross-area natives were included in this claim. Judge Goldstein told the villagers that a road would soon be built through the area and the claim was made for their protection. . . .

"The road was built. Judge Goldstein and the Native Legal Services were never heard from again in the Tanacross area. Other claims, of which four were made between 1950 and 1967, have also been ignored. . . ."

The first land-claims effort that got results was made by the Tlingit and Haida Indians in the southeast. It was probably inevitable that the Tlingits and Haidas would be the first Alaskan natives to beat the white man at his own game. They had the most materially sophisticated culture in Alaska before the white man's advent, and since the white occupation they have done better than any other native group in business and the professions. The Tlingits and Haidas, an anthropologist told me, "had the Protestant ethic before the Protestants ever came."

The first organization formed to look after the natives' own interests, the Alaskan Native Brotherhood, was founded in the southeast in 1912 and is still active, although it now functions chiefly as a fraternal society. The first native elected to the territorial legislature (in 1923) was a Tlingit lawyer named William Paul, who was also the first native lawyer. (Paul, who is still very much alive, remembers clearly that in 1920 "Natives weren't allowed to vote. They weren't allowed to go to the white man's school. Restaurants had signs, 'No dogs or natives allowed.' ")

Most of the land the Tlingits and Haidas owned had been incorporated by the federal government into the Tongass National Forest, which takes up most of southeastern Alaska. In order even to sue for return of their land, the Tlingits and Haidas needed a special act of Congress permitting them to do so. In 1935, after ten years of work by William Paul and at the urging of Alaska's delegate, Congress passed a law permitting the Indians to sue. The legal process dragged on for decades. In 1946, Congress established the Indian Claims Commission to settle any claims made by Indians against the government, and in 1959, the Claims Commission finally decided that southeastern Alaska would have belonged to the Tlingits and Haidas if the federal government hadn't created the Tongass National Forest, and that therefore the Indians were entitled to some compensation. The court also decided that some 2,600,000 acres which hadn't been included in the national forest still belonged to the Indians. Nine years later, in 1968, the court fixed the compensation at $7.5 million.

The Tlingits and Haidas weren't the only natives who tried to recover their ancestral land. The people in Tanacross, despite the absolute inattention given to their first formal claims, kept trying. The *Alaska Review of Business and Economic Conditions,* a publication of the University of Alaska's Institute of Social, Economic and Government Research, said in 1967, "The Tanacross Village claim, filed in the early 1940's, again in 1961, and a third time in 1963, illustrates the position of many of the smaller native groups.

" 'We the people of Tanacross Village do here now place our blanket claim for the land in this area. There are twenty-one families living in our village. No one in the village is employed year round, and only two men have been able to find part-time work this summer. Some of older people get aid but they still must get part of their food from this land to be able to survive.

" 'There are thirty-seven traplines, nine fish camps, twelve berry camps, and the complete area we have claimed is used for hunting caribou, moose, ducks and for trapping. Our first blanket claim was sent into the Bureau of Land Management in the early 1940's, but it seems no one has a record of this

An Arco oil rig on the North Slope *Atlantic Richfield Company*

Pipe for the TAPS pipe-
line stacked at Valdez

*Joe Rychetnik
for TIME*

The proposed TAPS
route is indicated on
this map of Alaska by
a broken line

Wide World Photos

The Brooks Range

*Joe Rychetnik
for TIME*

A herd of caribou crossing
the Koyiukuk River

Joe Rychetnik for TIME

Indian children outside
their home in Minto

Karen Pataki

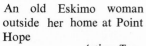

Second Avenue, Fairbanks

Karen Pataki

An old Eskimo woman
outside her home at Point
Hope

Arthur Tress

The Walter J. Hickel Highway, running 400 miles from Fairbanks
to the North Slope air oilfields   *Joe Rychetnik for TIME*

claim. So we the Tanacross Indian people do once again claim this land as ours.

" 'We have not had the opportunity to receive an education which would enable us to share equal employment openings, therefore we must have the land needed to at least be able to feed and clothe our families and to see our children gain the education they must have.' "

In the far north, the idea of claiming land grew slowly. Simon Paneak, the old Eskimo from Anaktuvuk Pass, recalls, "I first heard about land claims in 1925, when I was living over near Barter Island, trapping foxes. ["We were trapping white foxes and colored foxes, too," he remembers. "We needed furs. They were the only way we could get money. We needed money to buy groceries and tobacco. We were learning to eat the white man's food—not like our grand-fathers—and Eskimos like to smoke."] The person who started the talk about claims was a schoolteacher in Barrow, a white man. Nobody paid much attention. It wasn't until around 1950 in Barrow and Kotzebue that young men who had been educated got interested."

Perhaps because the land itself is so important and conspic-uous a part of Alaska, native land claims have attracted a lot of attention. E. L. Bartlett, then the Alaskan delegate to Con-gress (and later a Senator from Alaska), said in 1944, "No matter has come before the Alaska public in years which has occasioned so much discussion, public and private, and which has aroused such controversy."

The controversy didn't reach its height until after 1959, when Alaska became a state. The Statehood Act gave Alaska the right to select 103 million acres, but it also said specifi-cally, "The State and its people do agree and declare that they forever disclaim all right and title to . . . any lands or prop-erty (including fishing rights), the right or title to which may be held by any Indians, Eskimos or Aleuts. . . ." This provision of the Statehood Act followed the Organic Act of 1884, which established a civil government for what was then the territory of Alaska. "The Indians," the Organic Act said, ". . . shall not be disturbed in the possession of any lands actually in their use or occupancy or now claimed by them." The trouble was, neither the Organic Act nor the Statehood Act specified which land the natives used, occupied, or owned, and neither specified how native use, occupancy, or title could be legally defined or determined.

The state of Alaska wasn't bothered by those legal ambiguities or by any discernible scruples, so it simply ignored the natives' claims of ownership and began selecting its 103 million acres wherever it pleased Alaska's Department of Natural Resources to do so. The first choices included part of the Athabascan village of Minto and various other lands the natives had always thought were theirs. There was no reason to believe that the people of Minto, for instance, had any wish to part with their land. The *Alaska Review of Business and Economic Conditions* reported: "In 1951, the Minto Village Indians, along with several other native groups, filed a petition with the Department of the Interior asking for hearings to determine their land boundaries. The petitions asked that the lands they claimed be reserved under a section of the Townsite Act. The hearings were never granted, and, after statehood, a large portion of the land in question was included in an application for selection by the state. In 1962, the Bureau of Land Management director in Fairbanks dismissed the protest against state selection of land in the Minto area."

The state also selected part of the village of Tanacross, driving a surveying stake right in the middle of the village, and tried to sell parcels carved out of the village as "Wilderness Estates" at the New York World's Fair. It took the Prudhoe

Bay area, which no one but the Eskimos had ever used. It applied for the area around Point Lay, which was claimed by the Inupiat Eskimos.

Understandably, the Alaskan natives began to worry. After the state selections gathered steam, they organized a number of regional associations and filed a load of claims to the land. The Eskimos of the North Slope, who had already (in 1961) formed a cultural and quasi-political group called Inupiat Paitot (People's Heritage), in 1966 formed a group specifically to press their claims to the land between the Brooks Mountains and the Arctic Ocean. Charles Edwardsen, one of the organizers of the Arctic Slope Native Association, wrote to William Paul, the Tlingit lawyer, by then living and practicing in Seattle, to ask him to represent the group.

"We are in the process of organizing a Native Association composed of the Eskimo people on the northern slope of the Brooks Range in Alaska with the express intent of securing in court our aboriginal rights and title to said land," Edwardsen wrote.

"In that this land is presently being exploited for oil and other minerals by the State and Federal Governments and also by private agencies," he continued, "we on behalf of the people wish to be advised on the feasibility of obtaining an injunction in San Francisco Federal Court to halt further sale of leases and exploration, thereby forcing into court the issue of native land rights in Alaska.

"It can be factually shown that our ancestors have historically used this entire area for the past 5,000 years for hunting and fishing. It can also be shown that from time immemorial, they have made use of the pitch and coal of the area and that the mineral deposits of the lower slope of the Brooks Range were also known to them in that they used raw gold to make small toys and fishhooks for their own use.

"We wish to obtain an injunction on the ground that our people have not adequately been compensated for said land and that by decisions in similar cases, neither the Federal Government nor the State has a clear title to said land, and therefore cannot exploit such land until either compensation, adjustment, or title is settled by court decision.

"Further, we feel that an injunction could be obtained on behalf of a native individual of the area, but that it would be more effective if obtained by the association representing a major cross-section of the native inhabitants, and therefore are organizing this group to be known as the North Slope Native Association."

The North Slope natives never got an injunction, but through Paul they did file a claim to the entire North Slope. Paul also filed claims on behalf of the natives of Tanacross. All in all, the twenty-two native groups in Alaska have now filed claims to almost the entire state.

At the end of 1966, when the organizing of local groups and the filing of claims had acquired a lot of momentum, native leaders from all over the state got together and established the Alaska Federation of Natives. Emil Notti, president of the Federation, says that so far as he knows, what sparked the land-claims activity was the state's selection of land in Minto. Certainly, he says, that was what upset people in the Athabascan areas of the state. He himself just drifted, without ever planning to do so, into the thick of the land-claims fight. He then decided he wanted to talk with natives in other parts of Alaska who had gotten involved. "I wrote a few letters in April, 1969, to guys I had heard of but never met: people like Willie Hensley [an Eskimo from Kotzebue who was for a while executive director of the AFN], Jerome Trigg in Nome, Joe Upickson and Eben Hopson [the present executive director] in Barrow." He invited them all to meet with him in

Anchorage. "I expected fourteen or fifteen people," he says; "we had three hundred." At that meeting, the AFN was formed. The new Federation persuaded Alaska's two senators, E. L. Bartlett and Ernest Gruening, to introduce a bill that would have settled native land claims once and for all, giving the natives some 250 million acres. The bill got nowhere. But the federal government did act.

Despite the flood of native claims filed in 1966, it was obvious that if no one stopped the state from selecting land, much of the territory the natives claimed would be gone before the claims had a chance to be settled. It was also clear that the state's title to much of its land would be bitterly contested. To make sure the land wasn't gone by the time the natives' claims to it were settled—or, cynics say, to make sure that the state's title to the lands it selected was secure—Secretary of the Interior Stewart Udall in December, 1966, put a "freeze" on all federal land claimed by Alaska natives, saying that it couldn't be taken for any purpose except perhaps for roads, airports, or other "public facilities." A standard rhetorical explanation given by Alaskans is that Udall, a Democrat, had imposed the land freeze in order to punish Alaska for electing a Republican governor, Walter Hickel.

Certainly Hickel himself didn't take the land freeze lying down. Claiming that the freeze was illegal, the state promptly sued Udall in Alaska's U.S. District Court to make him lift it. The Alaska court ruled in favor of the state (on December 19, 1969, the Ninth Circuit Court of Appeals in San Francisco reversed the decision), but the freeze remained. Then Richard Nixon was elected President, and he made it known that the very same Walter Hickel was his choice for Secretary of the Interior. Udall, before he left office, signed Public Land Order 4582, which froze all federally owned land in Alaska until the end of 1970. What Udall had done Hickel could undo, but

Hickel had the bad taste to say so publicly before his appointment was confirmed. After several such slips of the tongue, the Senate Interior Committee turned Hickel's confirmation hearing into a long grilling, in the course of which the committee chairman, Senator Henry Jackson of Washington, got Hickel to promise that he wouldn't lift the land freeze without the consent of the Interior Committees of both the Senate and the House.

Hickel's object as governor had not been to defeat the natives but to end the land freeze and get the business of land claims cleared away. He not only sued Udall; he also tried to work out a settlement of the native claims. He held a conference of state and Alaska Federation of Natives officials in 1967 to work out a bill that would be satisfactory to both sides. The natives said they'd settle for eighty million acres. Hickel said that was too much and suggested ten million acres. They said that was too little. Finally, Hickel and the natives compromised on forty million acres. There was also the question of a cash payment to compensate the natives for the land they wouldn't get. Hickel got the Alaska state legislature to approve a state payment of $50 million as part of a settlement if the claims were settled and the land freeze was lifted by October, 1969. The bill proposed by the state and the natives got nowhere.

Other bills were introduced, too. In 1967, the Interior Department came up with a proposal that the natives' land be expropriated in return for payment at 1867 land values. In 1968, the Department recommended that the natives be given as much land and money as they needed to maintain themselves. President Johnson said in 1968 that the government should "give the native people of Alaska title to the lands they occupy and need to sustain their villages."

The second Interior Department proposal had a hearing

before the Senate Interior Committee in 1968. In the course of the hearing, it became painfully clear that no one even had any good idea of how many natives there were in Alaska, much less how much land they needed to sustain themselves or on what their claims to the land they occupied were based. Senator Jackson, the committee chairman, decided that it might be nice if someone found these things out. Alaskan Senator Bartlett urged him not to look outside Alaska for his information gatherer; the Federal Field Committee for Development Planning in Alaska already knew more than anyone else about the subject, Bartlett said, and should be allowed to do the job. Jackson agreed, and in a matter of months the committee, chaired by Joseph Fitzgerald, who is a former Rhodes Scholar and former president of Ozark Airlines, ground out a massive 565-page report, far and away the most thorough study ever done on Alaskan natives. It described the terrain in which they lived, their economic patterns, their history, and their current situation, concluding that "Alaska natives have a substantial claim upon all the lands of Alaska by virtue of aboriginal occupancy." The Federal Field Committee's report in effect legitimized native land claims.

Before the Committee issued its report, critics had found it easy to dismiss the native claims as based on nothing more substantial than opportunism and delusion. It is still possible to dismiss the claims, but it is no longer easy to do so, and dismissals carry little weight in Congress.

By the time the oil companies paid $900 million for North Slope leases in September, 1969, the natives had dug in for a long battle, and the stakes had risen considerably. As soon as oil was discovered on the North Slope, it became clear that the land over which the natives, the state, and the federal government were contending was worth a great deal of money. The natives considered applying leverage on the other parties concerned by getting an injunction against the lease sale, but they decided against it. Instead, they simply picketed the building in which the sale was being held, calling for settlement of their claims. They did, however, revise their estimate of how much they could reasonably demand as payment for the land they were giving up. In 1969 they presented a bill to Congress proposing that they be given clear title to forty million acres of land, a payment of $500 million in cash spread over nine years, and a perpetual 2-per-cent royalty, to be deducted from the state's 12½ per cent, on all oil taken from the North Slope.

Walter Hickel and the Nixon administration proposed 12.5 million acres, $500 million spread over twenty years, and no royalties. Both proposals went to Senator Jackson's Senate Interior Committee, and it was generally assumed that the Senate as a whole would more or less follow the Interior Committee's lead. Jackson said he wanted to get a bill out of committee by Christmas, 1969, and in mid-December his committee seemed to be moving toward some kind of decision. But then his sister was killed in a fire, he left Washington, D.C., for his family home in Everett, Washington, and the committee lost its momentum. It didn't get back to work on native claims until February, 1970, when it started deliberating behind closed doors in the hope of reaching a decision without pressure from the various interest groups involved in the conflict.

By the end of 1969, former Supreme Court Justice Arthur Goldberg had agreed to be the natives' chief counsel, former California Senator Thomas Kutchel had joined the legal staff, and former U.S. Attorney General Ramsey Clark (by then a law partner of Goldberg) had taken charge of relations with Congress. The AFN entered 1970 well represented—and insistent on the arbitrarily selected but increasingly symbolic formula of forty million acres, $500 million in nine years, and a 2-per-cent royalty, to be taken from the state's 12½ per cent. The administration showed no sign of wavering from its own proposal. The state of Alaska favored the administration's stand and in any case bitterly opposed giving up even one penny or one acre of land it regarded as rightfully its own. The state was in a hurry to obtain some kind of settlement so that the federal government could lift the land freeze and the state could continue selecting land and developing its economy. The natives were in a hurry to obtain a settlement because the land freeze was due to expire at the end of 1970, and once it expired, there would be nothing to prevent the state from taking land as it had before December, 1966.

The magic number of $900 million was important to both sides, too. To the natives, it symbolized how much they were giving up even if they kept forty million acres and how much white society could easily afford to pay. To the state, it symbolized how much there was to lose, and how much white Alaskans were fighting to keep. The oil companies, which had provided the $900 million, remained officially neutral. They just wanted to see everything settled so that they could go ahead and remove their oil and make their profits with minimal uncertainty and disturbance.

5

THE ALASKAN NATIVES' fight for land and compensation is hardly a great popular uprising. Although it might be hard to find natives who didn't know of the land-claims fight, and perhaps even to find natives who didn't hope somehow to benefit from it, the fight itself is being carried on by a small group. Even those natives in the villages who are interested in land claims receive information only sporadically and assume that the Alaska Federation of Natives is taking care of things. They view the efforts of the Federation from a distance, as they might view the efforts of the state or national government. Emil Notti, president of the Federation, acknowledges the villagers' passivity and their sketchy information but says that the situation can't be helped. Communications in Alaska aren't good, keeping in close touch with the scattered native population would be difficult and expensive, and the Federation has limited resources. "It costs $30,000 to send someone through all the villages one time," Notti says. "We've decided to concentrate our money and effort outside."

There's no reason to believe that the "average" native is more remote from the Federation than the average citizen of almost any country in the world is from his own government. However, governments have territories and power, and, in one way or another, hold their citizens very much as subjects. The Alaska Federation of Natives has no power and no territory, so verbal communication is the only real link between the villagers and the Federation leaders. To some Alaskan liberals, a surprising number of whom hold the Federation in low esteem, the weakness of that link makes the Federation's claim to represent the native people absurd at best. "The leaders," they say, "don't represent anyone."

But the representativeness of the leaders is almost beside the point. Congress, which has chosen to lump together as "natives" all the Alaskan ethnic groups despite their traditional mutual hostilities, has also chosen to regard the Federation as the natives' bargaining agent. And, in fact, however short of ideal the Federation's relationship with its constituents may be, there's no other organization to consider.

Some people who have been close to the Federation say that its leaders are ambitious. "Everyone is running for governor," I was told at the beginning of 1970. Even people who are sympathetic to the Federation and respect its leaders tend to think that the leaders are bent on doing well by doing good. Still, there's no reason to suspect that any of the AFN leaders are trying to advance themselves at the expense of the native population, or that the interested village native's hopes for a land-claims settlement differ radically from their own.

Both the younger leaders and the older village natives resent the prospect and the established fact of white civilization's taking over their lands. The traditional "reverence for the land" is strongest in the old village people. They have always lived with the land, hunted over it, known that they inherited it from their parents and grandparents; the prospect of outsiders' taking it over seems outrageous. Most of the younger leaders do not have this strong traditional bond with the land. They have moved out of the traditional culture, have grown familiar with the white man's cities and the white man's politics. Nevertheless, they believe the land is theirs, know it's valuable, and resent any attempt to take it away. The older village people tend to be incredulous that any outsider could and would take their land. The younger leaders are more intent on receiving some kind of compensation. Being more attuned to white society and more aware of the need to succeed in it, they place much more importance on the cash and royalties they have asked for as part of a settlement. Still, they are adamant about land. "At every board meeting, some lawyer says we should have a minimum figure for the amount of land we're willing to settle for," says Emil Notti. "He says it would make it easier for the people making up the bill if they knew the limits within which they had to work. So we go around the table. The first man says, 'Forty million acres, that's my minimum.' The second man says, 'Forty million acres.' That's the way it goes, all the way around."

Some of the natives' most influential white well-wishers wish they would think less about the land and more about the money. "Some of the older natives are hung up on land," says Joseph Fitzgerald. "They don't need much land. The important thing is to pump money in. If the capital isn't squandered, just a large cash payment will give them income in perpetuity." Senator Jackson, who is both chairman of the key Congressional committee and the natives' chief friend at court, substantially agrees. The officials of the Federation do not. When it's suggested that the possession of a huge amount of unproductive land might turn out to be less a blessing than a handicap, Emil Notti gets indignant. If land isn't a liability for white entrepreneurs and white-owned corporations, he asks, why should it be a liability for the natives?

Native leaders and white well-wishers also differ over how best to administer the money the natives get through either a straight cash payment or royalties. The cash settlement named in both the Federation plan and the government plan is $500 million, to be paid over a period of nine years in the first plan, twenty years in the second. Royalties, if any are granted, could total at least as much again. So the question is how to distribute and manage a sum that may be as large as a billion dollars. A lot of people in the villages would like the money to be distributed directly to the villages and to have each village manage its own share as it pleases. The Federation hasn't pressed for this solution. For one thing, even some people in the villages think that money distributed so widely to such small and unsophisticated places would find its way into too many of the wrong pockets. Jack Ahgook, the Federation delegate in Anaktuvuk Pass, thinks village corporations are a bad idea because "some people just want to get their hands on the money." Also, if the money is distributed among the villages, there may not be enough in any one place to finance any major economic development. Besides, no one seriously believes that Congress would hand over a billion dollars to a collection of ragged villages.

The Federation wants the money to be administered through twelve regional corporations, which the leaders feel would be more practical than village corporations and would respond more readily to the real needs of individual villages and villagers. A single statewide corporation they feel would be too bureaucratic, too remote, too much, as one native leader said, "like a new Bureau of Indian Affairs." And they point out that, after all, the members of the Alaska Federation of Natives aren't really a single homogeneous ethnic group; they comprise a number of different ethnic groups with different cultures, different languages, different needs, and, in many cases, a history of mutual hostility. The Eskimos and Athabascans, for instance, fought each other for centuries. Even now, there is friction between ethnic groups within the Federation. It is sometimes hard to get them to agree. And some ethnic groups have been known to join forces against other groups to get their own representatives elected to office.

Most of the natives' white allies, including both Joseph Fitzgerald and Senator Jackson, don't think much of the idea of regional corporations, to say nothing of village corporations, and believe that the money should be administered by a single statewide agency. The main argument in favor of a state corporation is simply organizational efficiency. "It's hard enough to find management for *one* corporation," Fitzgerald says, and he insists that twelve corporations simply won't work. Senator Jackson agrees. Both men think it simply isn't likely that the natives will come up with enough good or potentially good executives to run twelve corporations well. Also, it seems probable that Congress would feel most comfortable in giving so large a sum of money to a single, highly visible, substantial-seeming group. Alaska's Governor Keith Miller, whom the natives hardly consider an ally or even a friend, does favor the idea of regional corporations. He reasons that the shorter the chain of command between the administrators and the natives, the more likely individual natives are to actually see some of the money. Most cynical observers say, very credibly, that what concerns the governor most is the prospect that a single state corporation with assets of a billion dollars would have tremendous political power and hence independence, while twelve regional corporations would be more manageable.

The Federation leaders aren't eager to compromise on regional corporations or anything else. In part, they feel that to show any willingness to compromise would weaken their bargaining position—for just that reason, Emil Notti refuses to say how little land the natives would probably settle for. "People say we're not flexible," says Alfred Ketzler, president of the Athabascan Tanana Chiefs' Council and acting executive director of the Federation during early 1970. "They ask, what's your fall-back position? We don't have one. If we give a fall-back position, then that's what we'll get. We've already fallen back from eighty million acres to forty million. The people in the villages are pretty well set on forty million acres, $500 million, and 2 per cent. The land is the most important thing."

But what some outsiders feel is the natives' intransigence runs much deeper than tactics or the desire to drive a good bargain. Many natives seem to have invested the figures of forty million, 500 million and 2 per cent, all arrived at arbitrarily or through bargaining, with moral significance. A lobbyist at the Alaskan state legislature in Juneau told me, "The AFN has convinced itself that its proposal isn't a bargaining position but an absolute." The same man said that Willie Hensley, a state representative from Kotzebue who was then executive director of the Federation, "is less respected this year than last. He thinks his position is moral and anyone who opposes it is immoral. Therefore he tends to despise people who disagree with him."

The Federation leaders are indeed convinced that their position is morally right. The case seems very clear to them: their ancestors used and occupied the land, each ethnic group defending its territory against all invaders; when the white man came, their ancestors never formally relinquished possession; therefore, they still own the land and anyone who tries to take it without at least paying for it is trying to steal it from them. The land the natives once dominated is huge, and it contains what may be the largest oil field in the world, so they think that even a billion dollars, which comes out to less than three dollars an acre, would hardly be excessive payment. (Oil companies paid up to $28,000 an acre for land at the Sep-

tember, 1969, lease sale.) The natives' insistence on the morality of their position may make them difficult for some people
to deal with, but it hardly seems a point against them that
they are charged much more often with moral righteousness
than with cynicism.

The native leaders' intransigence is fed not only by their
conviction of their own moral rightness but also by a feeling
that both the state of Alaska and the oil companies are actively
trying to cheat them, and that within Alaska, they are under
bitter and unprovoked attack. "At first we tried to speak very
softly and not get anybody mad," Emil Notti says. "It didn't
work. There was a backlash anyway. So I don't worry about
it any more. To hell with them."

What the natives consider the main backlash occurred in
the fall of 1969 and began with an editorial in the state's
largest newspaper, the *Anchorage Daily Times*. On October
18, the same day the House Interior Committee's Subcommittee on Indian Affairs held a hearing in Anchorage on the
land-claims issue, the *Times* ran an editorial entitled "The
Goldberg Bill." The editorial said that the natives' proposed
2-per-cent royalty alone "would amount to hundreds of billions of dollars and is sufficient grounds to oppose the legislation. But that is only one of a basketful of fantastic proposals
advanced to the Congress by the native association and its
lawyers, headed by Arthur Goldberg of New York. . . . The
total effect [of the natives' proposed bill] would be to cripple
the development of Alaska for all its citizens. . . . This
Goldberg Line that has been fed into the Alaska Native
Federation bill is a threat to all the people of the 49th State."

At that afternoon's hearing, Representative Ed Edmondson
of Oklahoma called the editorial "journalism at its lowest
level" and said that "to put on [the] bill the label, 'The Goldberg Bill,' in the same editorial that makes an appeal for an

end to racial animosity is not only low-level journalism, it is gutter journalism, and I think it should be deplored."

In November, a group of Anchorage business men made a conference call to Alaskan Senator Ted Stevens and told him in effect that if he pushed for a generous land-claims settlement in Congress, he might not be Senator much longer. State officials did nothing to discourage anti-native feeling—many people feel that they actively encouraged it—and although no one burned any crosses, there was a good deal of public antagonism.

Federation leaders especially resent what they regard as the hostility of the state government. They believe that without the active encouragement of the state, public hostility might never have developed. "Hickel started the first backlash," Emil Notti says. "It was on February 7, 1967. He tied up every radio station in the state for half an hour with a speech in which he said, 'Just because somebody's grandfather chased a moose across the land doesn't mean he owns it.' " Notti feels that as governor, Hickel was devious as well as hostile. "I knew in December, 1968, that Hickel was going to be nominated for Secretary of the Interior," he says. "He wanted our support, so he called me to his house one night and said, 'We'll take care of you.' Then the state selected oil land right around native villages. That's how he took care of us. We went to Washington two weeks before his confirmation hearing and spent two weeks knocking on doors. We told people, 'He'll lift the land freeze.' Finally, the committee made him promise not to do it."

Needless to say, the Federation leaders aren't comforted by the fact that Hickel is now Secretary of the Interior. (They aren't very happy about the department over which he presides either. They think that almost everyone in the Interior Department who knew anything about their problems either left routinely when the Nixon administration came in or was actively purged. They feel that the Department doesn't understand their situation and isn't interested. They may be right. A high official in the Department told me contemptuously in January, 1970, that the discovery of oil on the North Slope had made the settlement of native land claims more difficult, because it had "given the natives visions of sugar plums.") But at least Hickel was willing to give the natives something, if not sugar plums, in order to get the land-claims issue settled. As governor, he did persuade the Alaska legislature to offer $50 million, and as Secretary of the Interior, he did get the Bureau of the Budget to approve $500 million. ("Wally Hickel doesn't give a *shit* about Indians," an attorney for the natives explained, "but he believes in getting a job done.")

The natives think that Hickel's successor, Keith Miller, and his administration have been stingy and hostile well beyond the bounds of intelligent self-interest. The natives say that no one in the state administration knows or cares anything about native problems, that the governor and the state have no consistent policies but are ruled by emotions and outside pressure, that they shamelessly stirred up a "white backlash" that could easily have been prevented.

Critics of the state believe it would have a better case in opposing the natives' demands if it had ever tried to do much for them in the past. But the state has done very little for the natives. In 1969, Alaska Legal Services filed suit on behalf of natives in the Fort Yukon region, who complained that the state had refused to deliver U.S. government food stamps. The state claimed that it didn't have enough money to distribute the stamps. Presumably, now that the state is oil-rich, it will be able at least to distribute stamps. But with all the talk of how to spend the oil money, the one point of substantial agreement has been that the money shouldn't be spent to help the natives.

This attitude is entirely appropriate in view of the way in which the state acquired the North Slope in the first place. "Aware that the North Slope was a potentially rich petroleum province, the state, in 1964, applied to the Federal Bureau of Land Management (BLM) for about two million acres lying along the Arctic coast between the Colville and Canning Rivers, an area which includes Prudhoe Bay," wrote Luther Carter in the October, 1969, issue of *Science*. "Although the region was part of the Eskimos' traditional hunting and fishing grounds, the state application said that it was free of aboriginal use and occupancy. BLM, for its part, published a legal notice in several issues of *Jessen's Weekly* of Fairbanks, a newspaper with a circulation of a few hundred in the villages of arctic Alaska, saying that any persons claiming land for which the state had applied should file an objection. No claimants appeared, and the land went to the state for it to dispose of as it chose, with or without regard for native interests. Commenting on this, Jane Bender, an Alaskan journalist, has suggested that, even though the procedure followed by BLM was legal, its morality was dubious. 'The burden of proof,' she wrote in an article . . . in the *Anchorage Daily News,* 'was placed upon people who could not be expected to untangle the legal phraseology, who might not even have seen the notice in the first place, and whose knowledge of the far-reaching consequences of that simple small-print notice might be said to be minute.' The state moved quickly to put up its North Slope land for competitive lease. . . ."

Giving natives proper notice has never been one of the state's specialties. When the state built a winter highway for hauling supplies to the North Slope, for instance, it put the road through the Brooks Mountains at the easiest and most logical point, Anaktuvuk Pass. "The state took a right-of-way right through our village," recalls Jack Ahgook. "Nobody ever told us about it. We woke up one morning and the state was here."

The Federation leaders expect the oil companies to be every bit as devious and stingy as the state. The oil companies claim that they will provide jobs for the natives, and that therefore it is to the natives' advantage to have oil fields developed and a pipeline built. Federation officials don't believe them. They say a token few natives have been hired and a few more probably will be hired in the future, but they doubt that natives will get good jobs or will get many jobs at all. TAPS has promised the Department of the Interior that "continuously during pipeline construction [it] shall conduct a pre-employment and on-the-job training program for Alaska natives adequate to qualify them for initial employment and for advancement to higher paying positions thereafter. [TAPS] shall do everything within its power to secure the employment of those Alaska natives who successfully complete [the] training program." Native leaders doubt the promise will produce many jobs. It had originally been proposed that TAPS promise to hire a certain quota of natives, but TAPS argued that there weren't enough qualified natives to fill a quota, so the requirement was dropped. Emil Notti claims that under the existing requirements, there's no reason to doubt that there will be outright discrimination. "The oil companies will have clean hands," Notti explains, "but they contract out all their work, and the contractors don't hire natives."

In their direct dealings with the oil companies, some natives already feel that they've been cheated. The Department of the Interior asked all six native groups that claimed land in the path of the pipeline to grant waivers explicitly consenting to the lifting of the land freeze and granting a right-of-way permit. Five of the six groups agreed. The sixth, the Arctic Slope Native Association, did not. The Eskimos' attorney, Frederick

Paul (a son of William Paul), opposed signing the waiver because he thought that by relinquishing their legal right to ob-
struct the pipeline, the natives would be giving up what little
leverage they had to compel a settlement of their claims. The
Eskimos weren't eager to sign anyway, and they didn't. The
Athabascan Indians of the Tanana area, around Fairbanks,
did sign. TAPS and oil-company representatives had spoken
with native leaders, including Alfred Ketzler, and convinced
them that if the natives signed waivers they could expect substantial economic gains: natives would be hired for the work,
and native companies would be given contracts if their prices
were even roughly competitive with white companies. The
natives expected the oil companies to come across—"You
take a man at his word," Ketzler later explained—but they received no contracts and almost no jobs. The native leaders
were outraged. "We've been cheated by a bunch of sharp oilmen," Ketzler said bitterly after it was clear that he and his
people would get nothing. But the natives didn't take their
bilking lying down. Instead, the Tanana chiefs got together and
decided to rescind the waivers and to sue TAPS for $40 mil-
lion in damages.

The brief for the suit, which was filed on February 4, told
the natives' story of how they had been conned. "Prior to July,
1969 [the original date by which TAPS had asked for a right-
of-way permit]," it said, "the defendants [TAPS and the pipeline companies of all the involved oil corporations] began
approaching the plaintiff villages [Allakaket, Bettles, Minto,
Rampart, and Stevens Village] . . . [to ask the villages for]
waivers or releases of their claims to the land over which the
pipeline will pass. . . . The defendants . . . [said] that if
releases of the native claims were executed, the plaintiffs
would receive preferential employment opportunities on pipeline construction work, and preferential treatment in obtaining

contracts on the pipeline construction. The . . . DNH Development Corporation was organized to take advantage of the promised contract rights. The defendants promised that the DNH Corporation would be given a 'right of first refusal' on contracts within its capacity; that it would receive contracts on a negotiated basis rather than on a 'bid' basis; that it would receive contracts upon which it was not the low bidder if its bids were 'competitive'; and that it would receive a 'letter of intent' for use in negotiating joint ventures and other arrangements with the third parties possessing technical or financial ability, for the purpose of taking advantage of DNH's preferential contract rights.

"At a meeting financed by the defendant on behalf of TAPS held on July 27, 1969, the Tanana Chiefs' Conference, a regional native organization acting on behalf of the plaintiff villages, executed a resolution authorizing construction of the pipeline in return for the promised employment and contract preferences. On the basis of the Tanana Chiefs' resolution . . . the plaintiff villages signed 'releases' offering to release their claims and agree to the pipeline construction in return for 'one dollar . . . and other considerations,' i.e., the promise of preferential employment and contract opportunities. . . .

"On September 11, 1969, the defendants met with representatives of the plaintiff villages. The plaintiffs delivered the signed resolution and release to the defendants and were given a letter of intent from TAPS. . . . The defendants repeated . . . that the plaintiffs would receive preferential employment and contract opportunities, that they would receive contracts on a negotiated basis, that they would be given a 'right of first refusal,' and that they would receive contracts upon which they were 'competitive' even if not the low bidder. . . . The representative of the defendants explained that the letter of intent didn't go that far because it would cause an adverse reaction among other contractors, that its purpose was use in negotiating with third parties, and that TAPS and its constituent companies should be 'trusted' to abide by the terms of the oral agreement. He also promised that in furtherance of the oral agreement, the DNH Corporation would be placed on the TAPS 'mailing list' for bid invitations and proposals. The one dollar was not paid.

"On September 15, 1969, the defendants filed copies of the resolution and release with the Bureau of Land Management in an attempt to induce the Department of the Interior to issue a pipeline right-of-way over the land affected. . . . On January 1, 1970, Secretary Hickel modified the land freeze as to the pipeline route. . . . The defendants applied to the Department of the Interior for a pipeline right-of-way on December 22 and December 29, 1969. . . .

"The plaintiff DNH Corporation has received no contracts from TAPS, despite numerous attempts to obtain them. It has not been given a right of first refusal on any contracts, nor given an opportunity to receive negotiated contracts. On numerous occasions DNH has requested that it be placed on the TAPS mailing list for invitations to bid and proposals, but the defendants have failed and refused to put it on any mailing lists. The defendants have frequently failed and refused to provide DNH with information or specifications necessary for bidding on jobs, and have frustrated attempts by DNH to obtain information and bid on jobs. Other contractors have received information, specifications, invitations to bid, and contracts from TAPS.

"Each of the plaintiff native villages has taken formal action to rescind any release which may have been granted to TAPS. . . ."

The day after it was decided to rescind the waivers, Alfred

Ketzler said in Fairbanks, "I feel a lot better since we made that decision. I went around to the villages and persuaded people to sign the waivers. I felt it was my responsibility." Not only would the natives rescind their waivers, Ketzler said, but "if they go ahead and build it anyway, we'll enjoin them."

The natives subsequently decided to beat TAPS and the government to the punch by asking for an injunction right away to prevent Secretary Hickel from issuing a permit for a right-of-way through lands they claimed. In March, the U.S. District Court in Washington, D.C., issued a temporary restraining order that lasted until April 1. Then, on April 2, federal Judge James Hart issued an injunction on behalf of Stevens Village. It seemed certain that TAPS would not be able to start construction before the spring of 1971, a full year behind schedule. ("It was a mistake to pick on the Athabascans," an Alaskan anthropologist told me. "Athabascans are *mean.*") For the first time, the natives had acted directly against the oil companies and against the whole course of North Slope development.

The natives have realized for years, though, that they had the power to impede Alaska's economic development in general and the development of North Slope oil production in particular. They talked seriously of enjoining the September, 1969, lease sale, and, although they decided not to try that, the state government is well aware that through the land freeze, native claims have already prevented economic development in most of Alaska. (The state administration is probably optimistic about the amount of development that would have taken place without the freeze.) The natives are frankly determined to keep the state tied up indefinitely if their claims aren't settled. They were heartened by the decision of the Ninth Circuit Court of Appeals reversing the state's earlier triumph in its suit to have the land freeze lifted. They believe that the decision imposes what Ramsey Clark has called a "judicial land freeze" on Alaska.

Some native leaders think they not only have the power to tie up Alaskan development but also form a decisive swing vote in Alaskan politics. The *Alaska Review of Business and Economic Conditions* reported in 1967 that one reason for the natives' assertiveness about land claims was their "growing awareness . . . of their political strength. Adult natives comprise about one fifth of the voting population, and 1966 was an election year. Political candidates spent extra days visiting the far-flung native villages, and their stance on the land-claims issue made headlines." The natives believe that the native vote, which is traditionally Democratic but in 1966 went Republican, is what gave the close gubernatorial election to Walter Hickel. Hickel reportedly isn't convinced, but native leaders now believe that they can make or break other political candidates in the future.

Although the native leaders believe firmly in their ability to exert political and legal pressure, they haven't relied on it. Their strongest appeal, both within Alaska and in Congress, has been moral. The Federation's most impressive piece of printed propaganda is an eighteen-page booklet entitled *Native Alaska: Deadline for Justice*. The introduction to the booklet reads:

"The United States and its people are offered a priceless opportunity to do justice to its aboriginal people whose treatment in the past has reflected little glory on our nation.

"A hundred years ago on the Western frontier, Indians and whites were killing each other for possession of the land. Today in Alaska, 60,000 Indians, Eskimos, and Aleuts are fighting to preserve their lands from expropriation by the state. They are waging a peaceful war for a decent share of America's future. Congress is now deciding their fate. The Alaska native people urgently appeal to the conscience of every American for help in their search for justice.

"Alaska's Indian, Eskimo, and Aleut citizens have conclusive legal and moral rights (original Indian title) to 340 million acres of land—90 per cent of the Alaskan landmass. They are asking Congress to grant them formal legal title to forty million acres essential to their present livelihood and future well-being, and for just compensation for the remaining 300 million acres they feel are beyond the possibility of saving. Their hopes are expressed in legislation submitted to Congress, and presently pending before the House and Senate Interior Committees.

"The decision Congress makes will profoundly affect the lives of Alaska natives for generations and will reflect on the honor of our nation for centuries."

The booklet proper then begins: "To the Alaska natives, the land is their life; to the state of Alaska, it is a commodity to be bought and sold. Alaska native families depend on the land and its waters for the food they eat, hunting and fishing as they have done for thousands of years."

Then comes a quote, set off in blue type: " 'Nothing is so sorrowful as for a hunter, empty handed, to be greeted by hungry children. They will look at your feet. If there's blood on your feet, they know you got a moose.' "

Proving that "the land is their life" and that "Alaska native families depend on the land and its resources" has been a crucial part of the natives' struggle. The native leaders have based both their legal and their moral cases on the argument that they own the land by virtue of "aboriginal use and occupancy" —i.e., that they have used and occupied the land "since time immemorial."

The natives "used and occupied the land to the extent that it was used and occupied by human beings," Ramsey Clark told the House Interior Committee's Subcommittee on Indian Affairs in August, 1969. Clark then suggested that the Committee let Federation official Willie Hensley tell how his own Eskimo family had used and occupied the land when he was a child.

"You understand," Hensley said, "that there are 200 or more villages scattered around the state, and once you move outside of the urban areas—that is, Anchorage, Fairbanks, Juneau, and a few others—you are really in Indian or native country.

"In these villages scattered along the coast and the rivers of Alaska, we make our living supplemented by work occasionally [outside the villages]. . . . Life has not changed a whole lot from the old days.

"For instance, when I was growing up, we lived outside of the village like a lot of people did then. . . . Our family . . . had several locations that we used at different times of the year, and I estimate we used probably 100 to 125 square miles in our subsistence, the hunting and fishing which was our livelihood.

"We had to camp along the coast where we fished during the spring and winter and also we moved upriver probably ten miles, where we hunted muskrat and duck and also fished in the rivers, and then we moved down to the coast again and fished for other types of animals and also hunted seals.

". . . We had to use a lot of land, and you compound this by thousands around the state and you see the extent of our use and occupancy."

The Eskimos are the native group that has retained the most of its original way of life, and they are also the group that roamed the greatest distances over the most desolate territory in order to subsist. Eskimos have therefore been used as the prime, or at least the most striking, example of aboriginal use and occupancy. A number of Eskimos prepared statements to read before the House Subcommittee on Indian Affairs in September, 1969. The younger people talked largely about how angry they were. The older ones described their memories of the old way of life. Weir Negovanna, James Kagak, and Samuel Agnassaga, three old people from the village of Wainwright, said:

"We want to tell you as the appointed delegates of the old people of Wainwright about the ways of our people.

"We lived off the land, going many miles up our rivers and over our lands to provide for our families.

"Up the Utokok River families had their own fishing and hunting camps where they would go at the proper time of the year. We would get our grizzly bear up in the mountains a hundred miles from the coast, caribou which we sometimes put in corrals, wolves, wolverine, all kinds of foxes, moose, lynx, squirrels, porcupine, elk, marmot, and mountain sheep.

"Mr. Hopson [Eben Hopson, then executive director of the Arctic Slope Native Association and now executive director of the Alaska Federation of Natives] tells us you do not believe that we travelled a hundred miles in all directions inland. You are wrong. All you got to do is look at our country to believe us. . . .

"From the sea, we got our polar bear, whales up to sixty or seventy feet long. Several kinds of whales we depended on like belugas, bullhead, killer whale, sperm, and others whose white man's name we don't know. . . .

"Near Point Lay we could not drink the water at a certain place because it was spoiled by oil seepage. At freeze-up time, we once in a while would set fire to hunks of oil. In our travels to relatives, we found oil up the Colville. We know it is oil because it smells like the oil we buy from the white man. Weir Negovanna states he once was low on oil for his gasoline motor, so he put some oil he found about fifty miles east of Barrow in his motor and he had no problems.

"The area near Wainwright is covered by coal, miles and miles of coal. We used it for heat and cooking. There are different kinds of coal, too. The government has been making the town of Wainwright buy a permit to mine the coal for about ten years now. We object to this, because it is our coal. . . ."

Another old Eskimo made this statement:

"My name is Lucy Smith Aḥvakana, an Eskimo born at Point Hope, but I was raised at Beechy Point, Alaska, just twenty-five miles from Prudhoe Bay. . . . There was no work in the northern area for the Eskimo in those times, only hunting and trapping for a living.

"I was happy with this kind of life because we knew no other way of life. We lived in a drift-log house covered with sod, and we used caribou skins for our bedding. We had seal-oil lamps for lights at night, also to burn in our stoves. In wintertime we looked for drift logs and wood washed up on the beaches from under blankets of snow for our firewood. Sometimes we went for miles with dogteams when we could not find it nearby to survive the extreme cold weather. This type of living cannot be compared with present-day living conditions.

"Men had to travel many miles, in fact hundreds of miles, in order to find food such as caribou and wild game with the help of dogteams. An average day's journey with a good team is twenty-five miles, and most of the time you could not stay on the sled, you had to run on foot to help your team along.

"When a hunter had more luck with his catch than another, he shared his catch with his neighbors. This is the Eskimo way of life. . . .

"I have a home at Beechy Point and it still stands. My parents, a brother, and my first husband are buried there. My children were born there. It was our home for many years.

"On my last visit two years ago I found I no longer could call it my home, because the white man had trespassed and taken over my land and home. The land around my home was torn, also the graves of my loved ones were trampled with machinery. I did not like what I saw and I went to see a lawyer in Anchorage. His name was Ted Stevens, a lawyer working for Mobile Oil at that time. [Stevens is now Alaska's senior senator.] I asked him in what way he could help me. The first question he asked me was, 'Do you have title to your land?' I answered him, 'I don't think I need title to my own land and home, because this was the homeland of my parents and their ancestors before them, for many years.' He did not say anything for a while, so I asked him another question: How come a white man can trespass onto my land with no permission, when he knows the Eskimos could not trespass onto white man's land? The lawyer's answer was that the white man did not know that the Eskimos lived way up in Beechy Point and offered to pay for the damage of my land in the amount of $2,500.

"Even though I was not rich I could not accept this offer because it left me nothing. . . ."

And 72-year-old Alfred Hopson said, "Research has found, digging through our old villages, that Barrow was inhabited 1,600 years ago and a nearby village called Birnik was inhabited 1,900 years ago. At 'Walapac,' twelve miles south from Barrow, . . . they found evidence from the old village to prove it was inhabited during the Punic age, 5,000 years ago. Our forefathers tell of travelling far and wide among settlements, how they fought Indians, suffering many massacres, but were able to keep the land. Many now living say the oil derricks on the Arctic Slope stand where their grandfathers fought and starved to keep their land."

The natives' arguments about aboriginal use and occupancy have left many people unconvinced. Many Alaskans believe that while the natives may have used and occupied most of the state in the past, they use and occupy relatively little of it today. And some people who concede that the natives still do use and occupy the land doubt that aboriginal use and occupancy will stand up in court as a basis of ownership. Senator Jackson believes that by legal precedent a claim based solely on aboriginal use and occupancy is clearly invalid. Even some of the natives' attorneys concede that use and occupancy alone may not be enough. When Representative E. Y. Berry of South Dakota asked Ramsey Clark, "What are the legal facts to support the contention that the natives owned all of Alaska?" Clark replied, "The law . . . is fairly clear that while use and occupancy is a measure of presence, the real question is dominion."

The natives believe that they can prove dominion, too. John Borbridge, general manager of the Tlingits and Haidas and first vice-president of the AFN, told the House Subcommittee on Indian Affairs:

"While we and the Eskimos and the Athabascans have combined here in the modern day, at a much earlier time . . . we asserted [the right of dominion] against one another, and our history is rich, not only the times that we came together but times that we frankly were asserting these rights as against one another, Athabascan against Eskimo and Athabascan against Tlingit. . . . In southeast Alaska where the Tlingits and Haidas asserted their dominion . . . our assertion of rights was over-all as a Tlingit and Haida group of people. We asserted this as against all other comers. . . ."

Representative Ed Edmondson of Oklahoma then asked, "When the Russians took possession of the area which they occupied, was there resistance to their occupancy and was there intent to exclude them from the place?"

Borbridge replied, "Mr. Congressman, I am very pleased that you asked that question. In Yakutat the Russians were —there may be a politer word than 'massacred' but we drove them out. In Amchitka, Alaska, likewise, we had an encounter in which we came out favorably. But perhaps even more importantly, when the Russians wanted fish or when they wanted game, they recognized that this was the land of the Tlingits and Haidas. They did not hunt on the land because it was not theirs, nor did they fish, but they traded with the Tlingits and Haidas so that we in effect were establishing our dominion as of that time and this was clearly recognized by them."

The native leaders are convinced that although a court case to establish the validity of their claims might take many years, they would certainly win. They cite the Tlingit-Haida case as their clearest precedent. The court's ruling that if the United States government hadn't taken land for the Tongass National Forest the land would still belong to the Tlingits and Haidas and the ruling that some two and a half million acres of land that weren't incorporated into the National Forest still belonged to the Tlingits and Haidas seem to the Federation leaders to prove that the natives still own all of Alaska that the federal government hasn't specifically appropriated.

Native leaders resent the fact that few people concede the merits of their legal case. Alfred Ketzler told me in early 1970 that he was tired of having people tell him, "You have a good *moral* claim, but not a *legal* claim." In July, 1969, Ketzler wrote a letter to the editor of the *New Republic* thanking him for a recent article on land claims but adding: "I want to correct Mr. Henninger [the author of the article, who had written, "The government's moral obligation to compensate the natives for their lost land rights is generally conceded, but establishing a precise legal obligation that could determine the terms of a settlement has proved virtually impossible."]: we have not lost our land, ours by law, and will not lose it until Congress expropriates [the land] . . . it is generally conceded that our title has not been "extinguished." Recognition of the natives' legal right to the land is important, Ketzler said, because "to those who view [the land claims] problem as one of alleviating poverty, and not involving property rights, our demand for some $10,000 and 800 acres per capita seems unreasonable, even outrageous." Recognition of "our legal rights," Ketzler said "is the keystone to a fair, generous and just settlement of our land claims."

Frederick Paul, the attorney for the Arctic Slope Native Association, made much the same point in an essay dated March 20, 1969. "Why is it," Paul wrote, "that if the sovereign wants to build a road over a white man's property, automatically one knows the sovereign must pay him its value; but when the sovereign wants Indian lands, one worries about the need of the Indians and justifies payment and the amount thereof by the criterion of need?

"President Johnson in a sense adopted this rule [when he said]: 'Give the native people of Alaska title to the lands they occupy and need to sustain their villages.'

"The 1968 Department of the Interior proposal [for settling native land claims], bearing Bureau of the Budget approval, was recommended to the Congress because of 'need.' . . .

"Why is it . . . that the amount of payment must be excused by 'what is necessary for the future economic and social development of the community,' as Walter J. Hickel, when governor, suggested? . . .

"How about the white man's rule of fair value?"

The natives haven't convinced Congress that their claims should be regarded in terms of white men's property rights; nor have they convinced even Senator Jackson, who has pushed harder for a settlement than anyone else in Congress, that their legal case is worth much. Still, few people in Congress or elsewhere doubt that on moral grounds, they deserve something. Most Alaskans think the natives should get some land and some money, and state officials would like very much to remove the legal barriers and threats to economic development. The prevailing feeling in the state is that the land-claims issue has dragged on much too long, and some kind of settlement would be welcome.

In Congress, the passage of a settlement bill depends more heavily on what Ramsey Clark concedes is the "fragile support" of individual conscience. Clark explains his own feeling about land claims by saying, "You have to go by your own experience, and mine is in the Southwest," where, he explains, the Indians have historically been pushed from bad land to worse, and the oil industry has taken petroleum out of the ground all around them, refined it and sold it far away, and left them with few jobs and little lasting benefit. Clark says that Congressmen from states with large Indian populations have seen the same or similar spectacles, and many of them would like to see the Alaskan natives do a little better. There is a feeling in Congress, Clark says, especially among the men from Indian states, that everything the government has done for American natives so far—the treaties, the reservations, the Bureau of Indian Affairs, the Indian Claims Commission —has failed dismally, and Alaska provides a last opportunity to do something right.

Clark told the Senate Interior Committee in August, 1969, that "Congress in a spirit of hopeful idealism created the

Indian Claims Commission, conceived to settle once and for all native claims, legal, equitable, and moral, against the all-usurping leviathan. The technique, an analogy to judicial process, was wrong, unrealistic, and harmful. It created conflict between the United States and Indian tribes—the conflict of massive and protracted litigation with its frustrations and injustice. It looked backward when the need was to look forward. The issues of fact were what happened often a century or more ago. It reopened old wounds, ancient wrongs, laid them bare, examined them minutely. The need was to uplift, to unite, to move ahead. It created fictitious issues. What was the fair market value of the sixty-eight million acres of land in California in 1853? But in 1853 there was no market, no ascertainable value, no buyer, no seller, and later developments known to all . . . had brought great cities, millions of people, gold and oil discoveries, orange groves, an agricultural product that would have fed the nation of 1853 twice over. It turned native people inward, toward themselves, set [them] apart and against their government and their fellow citizens. Finally, for something that happened to ancestors they never knew in times they never lived, they were paid dollars that were soon gone and left them more impoverished and embittered than ever." He says that Senator Jackson, who as a young representative introduced the bill that created the Indian Claims Commission, feels particularly responsible for that, and particularly eager to get a settlement through Congress.

The natives' main enemy in Congress, Clark believes, is just plain lethargy: native claims don't rank high on anyone's list of priorities. (Senator Jackson reported in late March, 1970, when his committee was supposed to be meeting every other day on the land-claims issue in an effort to get a settlement bill out, that simply assembling enough senators for a quorum was a major problem.) But some of the barriers to obtaining the settlement the natives want are much more specific. First, the amounts involved stagger the imaginations of many Congressmen. Forty million acres may be only about a tenth of Alaska, but it equals the total areas of several smaller states; $500 million may represent only three dollars an acre for land worth as much as $28,000 an acre, but it looks like a great deal of money to Congressmen trying to get a few thousand extra dollars for impoverished constituents of their own. Congressmen from states with large Indian populations fear that a generous Alaskan settlement would be regarded as a precedent and would stir up demands among the natives back home—this category reportedly includes such men as Senators Bellmon of Oklahoma, Anderson of New Mexico, and Fannin of Arizona, and even the generally liberal Senator Church of Idaho. Less widespread is the desire to protect vested interests in Alaska; Senator Mark Hatfield of Oregon, for instance, is reportedly opposed to any settlement that would give natives in southeastern Alaska ownership of timber lands currently exploited by the Portland-based Georgia-Pacific Company.

The biggest stumbling block in the natives' settlement plan is neither land nor money, though, but royalties. Ramsey Clark argues that "The royalties are the most important thing to put through. They represent power. They give the natives participation in Alaska's development. They aren't a dole." Jackson, too, thinks royalties are important. He is especially concerned with providing a continuous source of funds, so that a cash settlement won't be squandered in a fraction of one generation. (He is impressed by the fact that unlike most Indians in the "lower 48," the Alaskan natives haven't insisted on an immediate cash payment.) In the House Interior Committee, Representative Ed Edmondson of Oklahoma has pushed hard to give the natives a 2-per-cent royalty.

The opposition to the idea of royalties is formidable. The Interior Department argues that it does not favor a royalty for the natives' own good. The Department's reasoning is a masterpiece of logic. In February, 1970, Senator Jackson asked Walter Hickel for the Department's opinion of the natives' proposed land-claims settlement bill. Hickel replied, "From the standpoint of the Alaska natives it would be to their benefit to be entitled to a fixed amount of $500 million as opposed to an unknown overriding royalty. From the standpoint of the natives in Alaska, the $500 million cash settlement is, in our opinion, far superior to a settlement based on an unknown value. We are therefore opposed to the provision of [the bill] providing for a 2-per-cent overriding royalty *in addition to* the $500 million." Presumably the uncertainty of how much more than $500 million they were going to get would be too much for the natives to bear. At any rate, the administration's bill includes no royalties, and Jackson expects the administration to lobby against the natives' proposal.

Some congressmen just don't like the idea of royalties. Emil Notti says that when a group of Federation representatives went to see Senator Fannin the Senator said, "Boys, don't even talk to me about 2 per cent. I don't even want to hear about it." Also, of course, the state of Alaska believes it should get all 12½ per cent of the royalties and bitterly opposes any plan to give the natives some. Alaskans tend to feel that although the natives' ancestors may have deserved some kind of payment, the current natives don't. They're afraid that if the natives get a lot of land, they'll fence off areas that are currently open to the public for hunting and fishing. And they think the native leaders are "too pushy" and want too much. One Congressional staff member who has done research in Alaska says the white people there like to think that they got ahead on the frontier and they shouldn't have to help a bunch of shiftless natives. An Anchorage lawyer agrees that Alaskans have a "frontier mentality" and says, a bit more analytically, "Alaskans have never had much social conscience. I talked this morning to a [high school] civics class . . . about native land claims. There were thirty kids in the class. The attitude of the kids was, 'Fuck the natives, don't give them anything.' They say, 'Get out of Vietnam,' but they don't want to give the natives anything."

The natives will almost certainly get something, and the absolute amounts of land and money will undoubtedly be large, but there is no conceivable chance that they'll get all they're asking for. Even in the villages, Alfred Ketzler says, "people are a little apprehensive" about what the settlement will be, and native leaders who keep track of what's going on in Washington know full well that some compromise will be necessary. They will be lucky to get twenty million acres of land by the time a settlement bill clears the Senate. And they may clear the Senate with $500 million, but they probably won't make it through the House. The royalties are a question mark. If they get royalties at all, they won't get them "in perpetuity," which is what they want. Even Jackson thinks there should be a time limit on royalties. (Joseph Fitzgerald, who presided over the committee that wrote the report on "Alaska Natives and the Land," opposes a perpetual royalty on the ground that it would make racial separateness permanent. No one in a position to swing much influence or power besides the natives and their lawyers favors a perpetual royalty.) Then there is the problem of how the money will be administered.

It's pretty certain that the sooner a settlement bill is passed, the more the natives will get. When the land freeze is lifted, at the end of 1970, the economic pressure for a settlement will disappear. Perhaps more important, the further the September lease sale recedes into history, the less vividly people will remember that a small fraction of the state once rented for $900 million, and the more extravagant the natives' claims will seem.

Despite the urgency of getting a bill passed quickly, Ramsey Clark doesn't think the natives can afford to push Congress too hard. If they do, Clark says, "Congress will throw the settlement into the Court of Claims," which has never settled a case for more than $35 million and is simply not geared for big settlements. If the natives' case goes to the Court of Claims, Clark says, "it has already fallen flat."

6

On April 14, 1970, the Senate Interior Committee announced that it had worked out "a final settlement of all land claims of Alaska native people and villages." Senator Jackson said it was "a just and generous settlement which provides the essentials—money, land, a share in Alaska's mineral wealth and modern corporate institutions—to allow the Alaska native people to assume control of their own destiny and to deal with the urgent social, health and economic conditions they have faced for many years." Alaska's Senator Ted Stevens called the settlement "a series of compromises."

The bill proposed that the federal government pay the natives $500 million over a period of twelve years, that the natives receive 2 per cent of the royalties on Alaskan minerals located on public lands not patented to the state, and that they be given formal title to four million acres of land, with surface rights to three and one-half million more. The royalties would continue to be paid up to $500 million, making the total amount of the cash settlement one billion dollars.

The bill sets up village corporations to hold and administer the land, two urban native corporations—one for natives in Alaskan cities and one for natives living in cities in the rest of the United States—and three statewide organizations in

Alaska. The first, the Alaska Native Commission, would prepare a list of natives and native villages eligible to receive the
benefits of the settlement and would settle boundary questions
and disputes. It would consist of five members, at least two of
whom must be natives, appointed by the president and paid
by the federal government. The commission would function
for a maximum of seven years.

Of the two remaining organizations, the Alaska Native
Services and Development Corporation would distribute land
to villages and individuals, allocate funds to the urban corporations, administer the "leasable mineral estate of lands
granted under the bill" and the "lands granted for economic
potential and development," and review and pass on village
plans. The Alaskan Native Investment Corporation would
"handle investments and business for profit activities." At
first, the Services Corporation would get most of the money,
but after twelve years, the Investment Corporation would get
most. The Investment Corporation would have a twelve-man
board of directors, with four members selected by the natives,
five appointed by the president, and three members, who could
not be natives, elected by the other nine. The Services Corporation would have an eighteen-man board, with four members
appointed by the president and fourteen elected by the natives,
one from each of the twelve regions and two urban corporations. After twelve years, an Alaska Native Foundation would
be set up for "educational and charitable purposes" and would
receive 10 per cent of the Investment Corporation's stock.

Native leaders' first reaction was extreme disappointment at
the amount of land offered and considerable disappointment
at the absence of provisions for regional corporations. Indications were that they might try to kill the bill, but they decided,
at a meeting on April 20, simply to press for more land.

The Interior Committee was not eager to talk about the

discussions that had produced the bill, and there was no immediate indication of what the "series of compromises" to which Senator Stevens referred had been. It was obvious, though, that the committee had made an effort to avoid taking much money from the federal treasury and setting any potentially inconvenient precedents, and that the state of Alaska had been given a lot more of the burden than it would have liked. Along with the bill, the committee produced a comparison between it and the other proposed native land-claims settlement bills, pointing out that in the bill finally approved, "The cost to the federal government is substantially offset by the reduction in federal appropriations resulting from the transfer of Bureau of Indian Affairs programs and services to the State of Alaska and to general health, education, and welfare programs available to all citizens. . . . The Bureau of Indian Affairs program in Alaska currently costs in excess of $31 million a year." The committee also pointed out that "the bulk" of the royalty payments would come "from moneys which would otherwise be paid to the State of Alaska."

The state would seem to have paid for the prerogative to select land almost anywhere in Alaska—and also for its dreadful bargaining position. Alaska has virtually no political strength in Congress and is so isolated that, in general, an injury to Alaska is not an injury to anyone else. Moreover, the state receives privileged treatment from the federal government on congressional sufferance, so Congressmen from other states have it over a barrel. Alaska receives 90 per cent of the revenue from all minerals taken from public land within its borders—more than twice as much as any other state in the union. A handful of Western states gets 37½ per cent, and the others settle for 12½ per cent. Alaska receives 90 per cent because when the state was created, its economic prospects were so bleak that Congress figured it needed all the help it could get. Now the situation has clearly changed. A state that contains the largest known oil reserve in the Western Hemisphere might have a hard time justifying its right to what is essentially federal charity. Senator Jackson and advocates of the natives' cause were very much aware of this. They reasoned that Alaska would lose a lot more by giving up its 90 per cent cut than by letting the natives take 2 per cent, and that if Alaska insisted too much on being tight-fisted with the natives, Congress might well decide to be a little less generous with Alaska. It's entirely possible that this reasoning was used to persuade Alaska's two senators to back the settlement bill that emerged from the Interior Committee on April 14.

Besides, the state won't have to give up any royalties at all from the land it actually owns. This provision must have eased the minds of senators not only from Alaska but also from other states that have large Indian populations with lingering grievances. Those senators must also have been pleased by the amount of land the bill would provide. Giving 1 per cent of the land to a group that comprises 20 per cent of the population is hardly an extravagant precedent. Nor does it infringe much on Alaska's prerogative to select land of its own.

Any bill that settled native land claims once and for all would have definite advantages for Alaska. Senator Jackson said that the current bill "provides the means for a lifting of the land freeze and for going forward with the economic development of the state. It means that the cloud on land titles in Alaska will have been removed. It means that state selections under the Statehood Act may proceed free from the threat of litigation and conflicting claims."

For the natives, Jackson said, "the bill provides opportunity for a better life for themselves and for their children: opportunity for full ownership of their homes and the lands they have used for years; opportunity for a better education; opportunity to live a longer, fuller life, free from hunger, disease, and want; opportunity to enter business and the professions, to generate employment and contribute to the tax base of local communities; and, most important, an opportunity for individual identity and pride."

Jackson said the bill did something for the federal government, too, in that it "presents the Congress with the chance to provide the native people of Alaska with justice, hope, and opportunity and to end 100 years of less than benevolent wardship." The bill means, he said, "that the last chapter in the sad history of the United States' relations with the Alaska Indian, Eskimo, and Aleut people will have a just, generous, and honorable closing."

In terms of money, the bill is certainly far and away the best the natives could reasonably have expected. Five hundred million dollars over twelve years is much closer to what they wanted than to the administration proposal of $500 million over twenty years. The royalties weren't granted in perpetuity, as the natives had demanded, but there had never been a chance of getting perpetual royalties, and there had been a danger of getting no royalties at all. The lack of regional corporations, or actually of any but the most cursory provision for regional differences among the natives, caused a problem right away. The North Slope people were particularly indignant about their share: 138,000 acres and $14 million. The North Slope representatives argued that they needed more land than people of other regions for subsistence hunting, that they had aboriginally used and occupied fifty-five million acres, and that the wealth from their fifty-five million acres would be providing most of the royalties on which the settlement was based; they therefore claimed they deserved a larger share. The Federation decided to ask for more land for the North Slope, and for at least eleven million acres over-all.

Land in general was, of course, an inevitable source of discontent. No informed native leader had really expected to get the forty million acres the Federation was asking for, but they had regarded the administration's proposal of 12.5 million as the minimum. The word is that Jackson himself pushed hard for more land. Obviously, other members of his committee felt they had a strong vested interest in keeping the amount of land small. And the native leaders had figured, well before April 14, that Jackson was too good a politician to let a bill leave his committee without unanimous support, which would, of course, reduce the likelihood that the bill would be attacked on the Senate floor.

The native leaders had said that they wouldn't passively accept a settlement they considered too small. Emil Notti had insisted that if the natives lost on one part of the settlement they should be able to make up their losses on another part; if they got less land than they wanted, for instance, they should get more money. Notti said in February, 1970, in Tacoma, Washington, that if the settlement wasn't satisfactory, the natives might go to the United Nations or the World Court to try to get status as a separate nation. The day after his speech at Tacoma, Notti admitted candidly that he had said what he did because "on the way down I was trying to figure out what I could say that would get in the papers," but he also insisted that he was at least "half serious." He was pretty sure that if the natives wanted to go to the U.N. or to the Hague, they could at least cause the United States a lot of international embarrassment. And a trip to the Hague wasn't the only possibility: the violent and non-violent demonstrations that some people foresaw if a settlement was delayed could easily have been attempted by a group of natives who thought the settlement was too small.

When the chips were down, though, the native leaders realized that the Interior Committee's proposal was the best they were going to get. Killing the bill would gain them nothing. The native leaders had little choice. Their moral arguments were their strongest weapons all along; if they had been reduced to quibbling about amounts, those arguments would have lost their effectiveness. And insisting that a billion-dollar settlement was too small certainly wouldn't have won the natives many friends.

More significant than what the natives might have done to protest the settlement is what they will do with the land and money the settlement will provide. Right now, the Federation leaders have no real plans for the money, no concrete idea of what the natives can do with it. The money itself will come in slowly, and a lot of it will probably be invested. It seems certain that some of it will be used for education; some schools may be built, and scholarships will almost surely be given to young natives who want to attend college. Money may also go to native entrepreneurs who need loans to start their own businesses. And money will probably be used to start or help start industries, either full-scale commercial ventures or village crafts. As things stand now, Emil Notti says, "The Bureau of Indian Affairs comes in, says 'This town needs a cannery,' and builds a cannery. Maybe that's *not* what the village needs, but the BIA is detached from the realities of Alaska." James Thomas, the Federation's former director of public relations, who has run businesses of his own, says that even white Alaskans tend to look too simplistically at the problems of economic development. When the Brookings Institution held a series of seminars in Alaska to find out how Alaskans thought the state should spend its windfall oil money, Thomas says the prevailing thought seemed to be, "Just plunk an industry down here and plunk one down there and everything will be peachy. It's not that simple," he says. "You can't get people to change their way of living overnight. If people fish, you should help them to be better fishermen." Notti says that not only should industries be set up to fit local needs and inclinations, but the primary function of new industries should be to provide jobs, not to make profits. "If a factory just breaks even but provides several jobs, it's doing well," he says.

Not all natives will be prepared to take advantage of college educations or factory jobs, though, and not all native villages will be in a position to profit from the establishment of local industry. Notti concedes that some villages probably can't be helped economically and that some natives will profit only marginally from even the largest conceivable settlement. But the settlement could make the transition to a modern white culture easier for the natives as a group, and if it includes enough land, it should enable the older people, those firmly rooted in the old cultures, to live out their lives in the old ways.

To say that there may be no economic future for some native villages and that the days of the traditional native culture are surely numbered is not to say that the villages will disappear in a year or two and the villagers will therefore lose their need for the surrounding land. The standard theory is that the villages *are* disappearing. The state government believes they are, and the oil industry believes they are, or at least that they will. Robert O. Anderson told me that he thinks the effect of economic development, oil-industry money in the pockets of some native villagers, will be an exodus from the villages and the disappearance of native culture.

This theory and its variations are entirely plausible, and as long-range projections they're almost certainly accurate, but for the short range, they are above all convenient. In December, 1969, at a Northwestern businessmen's conference on North Slope oil, I heard Frank Murkowski, then Alaska's Commissioner of Economic Development, give the audience a quick rundown on the land-claims issue. As Murkowski covered point after point, he read off three sets of figures: the federal government's, the state government's, and the natives'. On every point but one, the differences among the three sets of figures were pretty consistent, with the federal estimate midway between the natives' and the state's. On the question of how many native villages still existed, though, the state's figure was much lower than the other two. This does not reflect a difference in anthropological theory or in the ability to count. Any native village that doesn't legally exist is a native village that has no claim on the land. Any native village too small to be counted is a native village that can safely be ignored. Emil Notti told me that when state surveyors went into the village of Tanacross, they told the people there that a village needed twenty-five families in order to be recognized (that was the figure in a land-claims bill pending at the time), and Tanacross had only nineteen, so Tanacross was simply out of luck.

To the state, to the oil companies, and to anyone else with a large vested interest in using the Alaskan land, it probably seems ridiculous to have large areas tied up by moldering villages of nineteen families. A full tribe of Hollywood Indians in beads and feathers, or a band of traditional Eskimos with their dogs and igloos, a perfect little museum exhibit, might be a different story. But the natives who remain aren't in museums; they wear store-bought clothing, hunt with rifles—which frighten seals away more quickly than harpoons would and therefore make it more difficult to kill many seals from one group—and ride on snowmobiles—which break up caribou herds, but do make it possible to run fifty-mile traplines. And, inconveniently, their villages aren't disappearing. Arthur Hippler, an anthropologist at the University of Alaska's Institute of Social, Economic and Government Research, says that the villages show every sign of surviving for a long time. Even if large numbers of young people go off to the cities, he says, the birth rate is so very high (and the death rate, thanks to the very recent availability of modern medicine, is low enough) that the present populations will probably increase. Those little villages will last long enough to use—and to demand— their ancestral land. To speak nobly of doing justice to the natives is really to speak of giving large parcels of valuable land to small, ramshackle villages that have no justification for existing in a modern economy. If the natives have a right to the land, they have a right simply because they are there, and because years ago, white settlers and entrepreneurs lacked the foresight to cheat their ancestors out of it formally. Notti concedes that a settlement inevitably will speed acculturation. It would be nice if some people could go on living as the natives used to, he says, but "there's no turning back."

The settlement will undoubtedly give the natives more political power and more friends among the same Alaska Chamber of Commerce types who have lobbied hard against them. It will help to make some individual natives wealthy and powerful and will increase politicking within and among native groups. It may give natives a bit more to be proud of and may lift the natives as a group a bit above the rock-bottom of Alaskan society. Whatever other benefits may or may not accrue, it will almost surely be easier for bright, ambitious native kids to go to college and for native entrepreneurs to start their own businesses.

But settlement money won't make the natives filthy rich overnight any more than oil money will make the state rich. As a way of showing that despite the large figures involved, the natives' claims were really very modest, Ramsey Clark pointed out several times that $500 million distributed over nine years still wouldn't raise the natives' annual income to half that of white Alaskans. (By the most generous estimates, the average annual per-capita income of Alaskan natives is about $600; that of white Alaskans is at least $3,600.) Five hundred million dollars spread over twelve years plus another $500 million spread over a period that may be much longer certainly won't produce a millennium. It will, however, give the natives as a group something very substantial to build on. And although in moral terms it can hardly atone for the United States' usurpation and destruction of the natives' land or the wrecking of the natives' traditional cultures, it may help to make their future slightly less of a national disgrace than their present or their recent past.

Inevitably, the rhetoric surrounding the settlement dwells heavily on "justice," but is justice really the question when the original owners of 375 million acres ask for forty million acres and are given four? Is it just, either legally or morally, to pay those same people one billion dollars for the remaining 371 million acres, when 10-year leases to just 450,000 of those acres have already been sold for $900 million? Of course not. Obviously, neither the United States nor any other nation would seriously consider giving 375 million acres of land, including many military bases and the largest oil field in North America, to 60,000 Eskimos, Indians, and Aleuts. The native leaders and their attorneys recognized that, which is why they didn't even *ask* for the whole state. But if the natives had gotten forty million acres, there would still have been 335 million acres left for everyone else. The United States, which has ruled the land since 1867 and has never used most of it, could certainly have afforded that. The state of Alaska, which has the right to select 103 million acres of its own, exclusive of the land the natives receive, could have afforded it, too. But greed and vested interests made forty million acres seem much too large. The state wanted to be able to select the best land, the most useful locations, the richest drilling and mining sites. Congressmen from states with large Indian populations didn't want the United States to decide formally that Indians' aboriginal right to the land took precedence over whites' vested interests. So forty million acres was impossible, for reasons that have little to do with justice.

The proposed settlement isn't devoid of justice, but the justice involved consists partly of recognizing the injustice of taking the immense riches of a huge, wealthy land and leaving the inhabitants of that land in abject poverty. The settlement embodies that recognition. It is essentially an effort to do *social* justice to a downtrodden group in a wealthy country, not to do *historical* justice to a people whose claim to their land is still legally or morally valid.

As may be inevitable when one attempts to do justice of that kind to people who are essentially powerless, the proposed settlement has a little of the "white man's burden" about it. The natives wanted land; their white well-wishers thought money was more important than land. They'll get money. The natives insisted they could manage a large amount of money by themselves; their white supporters were afraid they couldn't. The investment corporation will be run by a majority of non-natives. This is not to suggest that the proposed settlement is somehow shabby. Even if one regards it as just a gesture, a billion-dollar gesture is pretty grand. But the rhetoric speaks of justice, and the history books may speak of justice, and if some Alaskan natives don't see it that way, it may be well to remember that *their* rhetoric had stressed justice of a different kind.

7

WHEN IT WAS ASKED to grant a right-of-way through the last real wilderness in the United States for the largest private construction project in history, the Nixon Administration, all puffed up like a passenger pigeon with its concern for the environment, never questioned the project's basic desirability and never seriously weighed the possible benefits against the possible costs. The builders' *methods* have been examined and regulated, but there was never any doubt that the project itself would go through.

The project in question—or, rather, not in question—is a four-foot-diameter pipeline that the Atlantic Richfield, Humble, and British Petroleum companies want to build from the oil fields on the North Slope to the seaport of Valdez, 800 miles to the south. In October, 1968, Arco, Humble, and BP formed the Trans-Alaska Pipeline System (TAPS) to plan and build the pipeline. Almost the entire route ran across federally owned land, so on June 16, 1969, TAPS applied to the Department of the Interior for a right-of-way permit. TAPS asked for action on its application by July. The companies wanted to start building the pipeline in March, 1970, and they had to sign agreements with the contractors who would do the actual construction. They had already ordered 800 miles of forty-eight-inch steel pipe from Japan.

Under normal circumstances, the application might have been approved as a matter of routine. But circumstances weren't normal. For one thing, all federally owned lands in Alaska were locked in Stewart Udall's "land freeze." A pipe-line right-of-way couldn't be granted without lifting the freeze, which would inevitably involve scrutiny by two congressional committees and a good deal of publicity. The right-of-way couldn't be granted as a matter of routine.

And the TAPS application received very special treatment. In April, Hickel had set up a task force led by Russell Train, then Under Secretary of the Interior, to study North Slope development, and in June, at President Nixon's request, Hickel had enlarged the task force to include representatives from the departments of Transportation, Defense, Commerce, HEW, and Housing and Urban Development. The task force spent most of its time considering the pipeline. Senator Jackson's committee held hearings on the pipeline. The House committee held a hearing. In September, as a result of the task force's work, the Interior Department came out with a list of "stipulations" that TAPS had to agree to before any right-of-way permit could be granted. TAPS may not have been happy, but it signed. Senate and House committees finally gave their approval in October, but then the Interior Department delayed its approval even longer.

Needless to say, no one went to all this trouble exclusively or even primarily in the interests of a few Eskimos, Indians, and Aleuts. In fact, neither the Interior Department nor the congressional committees nor, for that matter, most of the natives themselves thought the pipeline would have much effect on the settlement of native land claims. But some people were very much concerned about the pipeline's effect on the land itself.

"The Brooks Range and North Slope of Alaska are accurately regarded as the last . . . wilderness in North America," said Train in a preliminary report on his task force's work to President Nixon. Those areas constitute "one of the few remaining major ecosystems left on the earth in a relatively unspoiled condition," said three conservationist groups —the Wilderness Society, Friends of the Earth, and the Environmental Defense Fund—in a suit to keep Walter Hickel from granting TAPS a permit. "No other area in the United States is as free of man's influence and environmental degradation," they said. John P. Milton, who in 1967 walked across the Brooks Range and the North Slope to the Arctic Ocean, one of the first white men ever to do so, wrote in *Natural History* in May, 1969, that the Brooks Range is "wilderness on a scale that the mountain men knew in our Far West during earlier days [with] . . . hundreds of miles of empty land and large expanses of unexplored territory. This Brooks Range wilderness still has these elements in abundance. . . . Here is an atmosphere of nature at its untamed, uncivilized best. The wilderness stands on its own: free, not propped by access roads, park rangers, interpretive centers and regulations on use, as in the quasi-wilderness of our national parks. Here there is no prostitution of the freedom so essential to wilderness—and the quality of the experience reflects this."

Conservationists were afraid—were, in fact, convinced— that the oil companies were about to ruin the entire area. As precedents, they cited the Navy's drilling in Naval Petroleum Reserve Number Four, just west of Prudhoe Bay, where oil drums and Caterpillar tracks left in 1944 are still clearly visible, and they cited other oil exploration in Alaska which has left piles of oil cans and refuse in some of the wildest areas of the state. John P. Milton wrote that during his walk through the Arctic, on August 19, "We climbed up to a long, narrow lake just west of the Jago River. An oil exploration crew had been there before us. The beauty of this lake is now diminished by the presence of over twenty scattered oil drums, the tattered ruins of a large domed shelter, and litter of all kinds strewn about. These scars will remain for a generation or more."

And then there was the first drilling activity around Prudhoe Bay, in the course of which gravel used for construction was ripped out of river beds, permanently altering the rivers. One road was built so badly that it eroded into a huge ditch, which in turn drained a major lake. By the spring of 1969, "Thousands of tons of gravel were removed from North Slope streams and coastal beaches to serve as a base for roads, airstrips, drilling rigs and buildings," Wilbur Mills wrote in a December, 1969, issue of the underground Seattle *Helix*. "Beaches were turned to mud, river channels blocked and silted, spawning areas of migratory fish destroyed.

"Miles of Cat tracks and seismic trails criss-crossed the area, resulting in thawing and erosion of the frozen tundra," Mills wrote. "In places the ruts were deep enough to hide a man. Garbage and debris were left to rot in the cold, dry climate, where it takes fifty years for a tin can to turn to dust. Frozen lakes were popular sites for winter camps. In spring the sites were not hard to find, the melting ice being well marked with garbage, fifty-five-gallon drums, and often-raw human sewage."

The oil companies have pointed repeatedly to Arco's drilling on the Kenai Peninsula, within the Kenai Moose Range, as proof that they can drill without destroying the environment. It has been done so carefully that the number of moose has even increased. But balanced against Kenai is the offshore drilling in Cook Inlet, where there has already been almost twice as much oil spilled as there was at Santa Barbara, and where for a long time every kind of refuse produced on the drilling platforms was simply thrown into the water. The Department of the Interior announced in a press release on April 1, 1968, that Secretary of the Interior Stewart Udall had just "called upon the Western Oil and Gas Association to join in a cooperative emergency control program and strict industry guidelines to cope with an increasing number of incidents of water pollution caused by oil exploration activities in Alaska's Cook Inlet.

" 'During recent months,' Secretary Udall said, 'I have received well-substantiated evidence that exploration and development activities in Cook Inlet have resulted in a recurring series of pollution incidents. Between June, 1966, and December, 1967, there were some seventy-five incidents of oil pollution in Cook Inlet reported by federal and state agencies responsible for the conservation of the natural resources of the area.'

"He said that pollutants have included crude oil, mud sacks, garbage, refuse, engine oil, stove oil, and jet fuel. While indicating that some progress has been made in talks with industry officials, he said the basic problem remains.

"As examples, Secretary Udall cited damage sustained in commercial fishing with oil fouling nets and fish taken in nets. In addition, there have been from 1,800 to 2,000 ducks killed by one oil spill alone. He reported that in December a tanker colliding with a dock caused an oil spill over a 20-mile area involving more than 1,000 barrels of oil."

To allow this same oil industry to build an 800-mile-long, four-foot-wide pipeline through the Arctic, critics have argued, would be to set the stage for an ecological disaster that would make Santa Barbara look small-time. The ecology of the Arctic is so fragile, they say, that even a slight disturbance could have far-reaching and disastrous effects—and the disturbance caused by construction of an 800-mile pipeline would not be slight. "In the extreme but relatively stable and regular conditions of the Arctic," wrote Barry Weisberg in the January, 1970, issue of *Ramparts,* "The web of life-supporting relationships depends on the slimmest margins of sustenance. The slender food chains and parsimonious life-cycles afford little tolerance for disruptions in the pattern of balance. The slightest manipulation of the life-support system, the alteration of a bird migration, the pollution of a river, the noise of an airplane, all have incalculable unanticipated consequences. That is what makes this unique and irreplaceable ecosystem so utterly fragile and so vulnerable to the careless intrusions of industrial man."

Conservationists have also argued that the pipeline, with a foot of insulation around it, would form an impassable six-foot barrier across Alaska, interrupting the migrations of the region's 400,000 caribou and threatening them—and the natives who depend on them—with extinction.

The U.S. Fish and Wildlife Service reported, "The fauna of the Arctic region is characterized by relatively few highly adapted resident species which are augmented periodically by many migratory species. The most prominent and important land mammal is the barren-ground caribou.

"The two largest caribou herds in Alaska are located in the Arctic region: the so-called Arctic herd (about 300,000 animals) and the porcupine herd (about 140,000 animals). The latter herd regularly moves into Canada. . . . The range of this herd in Alaska is generally within the Arctic National Wildlife Refuge. The range of the Arctic herd encompasses the western half of the state within the Arctic Circle.

"South of the Arctic area and within the area of pipeline

alignment consideration are two smaller herds. The Nelchina herd of approximately 70,000 animals ranges throughout the Upper Copper River drainage. . . . The Mentasta herd, estimated at 5,000 animals, ranges into the northeastern part of the Copper River area. . . .

"In the Arctic region, caribou normally spend the winters in the lower mountain areas and then in May move to the North Slope, where they bear calves. Later they move into the mountains for the remainder of the summer. The subarctic herds seasonally migrate between the river valley and high mountain areas.

"As is obvious, a forty-eight-inch pipeline placed on the surface of the flat terrain of the North Slope would form a fence or barrier where it crossed migration routes. . . ."

The oil companies' current plans, though, call for all but about fifty miles of the pipe to be buried, with no section longer than eight miles above ground. Even if, as now seems possible, the companies have been overoptimistic about their ability to handle the permafrost and are forced to lay substantial lengths of the pipe above ground, the caribou will probably be able to move back and forth across the line. If the pipe is on the surface of the ground, the U.S. Fish and Wildlife Service reported, "The detailed pipeline design will provide for continued migration routes either by raising sections of the pipeline to allow underpassage of the animals or [by] construction of ramps to afford overpassage."

But critics have pointed out that the pipeline would be more than a barrier, and that burying the pipe also involves hazards. The oil in the pipe would come out of the ground hot, and its heat would be maintained at 158° to 176° by pressure and friction all the way to Valdez. Whether the pipe was laid in or on the permafrost, how could oil at 158–176° avoid melting it? If the permafrost melted, the result would be ugly erosion at best. At worst, the pipeline would be undermined and perhaps broken. And if the pipeline ever broke, the oil would spread for miles, destroying every living thing in its path.

A few years ago, the federal government might not have lost much sleep over the possibility that a huge commercial venture by a politically powerful industry would make life difficult for caribou and tundra. The oil industry has long had so much political pull that, according to Erwin Knoll in the March, 1970, *New York Times Magazine,* "Some critics . . . describe the oil industry as 'the fourth branch of government.' . . . Under the benign patronage of such influential figures as the late Senator Robert Kerr of Oklahoma, who rejoiced in being known as 'the uncrowned king of the Senate'; the late House Speaker Sam Rayburn of Texas; the late Senate Minority Leader, Everett McKinley Dirksen of Illinois; and former President Johnson—all of whom shared a profound and undisguised commitment to the industry's welfare— the petroleum producers enjoyed decades of virtually limitless power in Washington. Their strength probably still surpasses that of any other special-interest group. . . .

"California oilmen were prominent contributors to the Nixon personal-expense fund that erupted into headlines during the 1952 presidential campaign," Knoll went on. "In Congress, Mr. Nixon was a reliable supporter of such oil measures as the tidelands bill, which divested the federal government of the offshore petroleum reserves. As Vice President, Nixon worked closely with Senate Majority Leader Lyndon Johnson in 1956 to block a sweeping inquiry into disclosures by the late Senator Francis Case of South Dakota that he had been offered a $2,500 bribe for his vote in behalf of a bill to exempt natural-gas producers from federal regulation. The law firm with which Nixon was associated before his 1968 candidacy had its share of oil clients. And oilmen—including President

[sic] Robert O. Anderson of rapidly growing Atlantic Rich-field—ranked high among contributors to Nixon's presidential campaign." Anderson often receives credit for Walter Hickel's installation as Secretary of the Interior. And Hickel, as a businessman and as governor of Alaska, had oil connections of his own. He was one of seven businessmen who in 1959 jointly obtained a franchise to distribute natural gas from the Kenai oil field in the Anchorage area, and he retained his interest in that enterprise after it merged with a pipeline company, not selling out until after his nomination as Interior Secretary.

But by the time TAPS applied for a right-of-way permit, the federal government's favors to oilmen had come under so much outraged national scrutiny and such strong attack that Congress was about to reduce the oil-depletion allowance, albeit only from 27½ per cent to 22 per cent, and—more important—ecology was well on its way to becoming a major national issue. The press gave ecology its standard big-discovery treatment at the beginning of 1970. On January 26, 1970, an article in *Newsweek* said, "Old Washington hands have been sensing for some time that environment may well be the key issue of the '70's, for the nation and for their political futures. They freely concede that no other cause has moved so swiftly from the grass roots into the arena of public policy-making. As early as 1968, environment was gaining on Vietnam in total linage in the *Congressional Record*. And by now, nearly everyone on Capitol Hill seems to be actively against pollution, causing a veritable stampede for stage center in the crusade to save America's land, air, and water."

The Nixon Administration was acutely aware of the public's growing ecological concern: the administration had come to power just in time for the nation's first big ecological scandal, the oil spill at Santa Barbara. Pictures of dead or dying seabirds, coated with thick black oil, made their way into newspapers, magazines, and television news broadcasts across the country. Around Santa Barbara itself, even people who had hitherto been politically conservative began shouting for Congress to clamp severe restrictions on the oil industry. The political pressure that built up was felt very keenly in Washington. "Secretary Hickel . . . is 'special interest' personified," wrote Steven V. Roberts, in the March 15, 1969, issue of *New Republic,* "and yet public opinion apparently can make a difference. Hickel has surprisingly become an advocate of stringent controls on oil exploration. . . ."

It must have been obvious to even the duller political minds in Washington that association with—to say nothing of responsibility for—an ecological disaster that dwarfed the one at Santa Barbara wouldn't do the administration much political good. Even with the worst will in the world, few people have denied that Nixon is politically astute or that Hickel is ambitious and shrewd. So the government was very definitely interested in minimizing the ecological damage caused by the pipeline, or at least in demonstrating its eagerness to do so. The task force spent most of its time considering the pipeline's impact on the environment. The congressional committees did the same. And the stipulations dealt almost exclusively with ways of minimizing environmental damage. ("I'd like to think that we would have done what we did if Santa Barbara had never happened," says former Under Secretary of the Interior Russell Train, but he has his doubts. Train says that Santa Barbara has had a tremendous effect on federal decision-making.)

The oil companies predictably tried to show that, far from wanting to despoil the Arctic environment, they were actually taking great pains to protect it. The state of Alaska, which was as eager as the companies to see the oil start flowing out, was also as eager to prove that the building of an 800-mile pipeline would be ecologically harmless. The state people had a particularly touching argument. Alaskans themselves, they said, knew better than any outsiders, even outside experts (a traditionally mistrusted class in Alaska), how to preserve the state. "Alaskans know Alaska," said Senator Ted Stevens at the Alaska Science Conference in August, 1969. "I'm fed up to here with people who try to tell us how to develop our country."

Alaskans probably have a little more justification than people in the other forty-nine states for feeling that they know their land. State Commissioner of Economic Development Frank J. Murkowski told the Senate Interior Committee in October, 1969, "We like the wilderness frontier . . . at this particular time last week I was out duck hunting with the Commissioner of Natural Resources. There were few birds flying. The sun was so warm that we decided to take a swim. My point being that we live [in Alaska] because we enjoy the outdoors. In fact, I think . . . that for the most part Alaskans, with very few exceptions, are conservationists." Murkowski's nature reminiscences were quite relevant. A lot of Alaskans do hunt and camp and fish, and a lot of people do live there because they like the woods and mountains and wildlife. But of course all across the "lower 48" states, "from sea to shining sea," people have been drawn to the wilderness and yet have systematically destroyed it for profit or convenience or just because it was there. There's no reason to believe that Alaskans are any less destructive or appreciably more careful or high-minded in their approach. State conservation officials have told me that Alaska is making exactly the same mistakes made earlier in the lower 48. "In the state of Alaska," Charles Konigsberg, professor of political science at Alaska Methodist University, said in October, 1969, ". . . the most powerful economic and political forces are afflicted with what has aptly been described as 'development psychosis.' " Not only do "we like the wilderness frontier," Murkowski said; "we intend to preserve that wilderness *in a manner that is consistent with the development of Alaska resources.*" State Commissioner of Natural Resources Thomas Kelly, Murkowski's swimming companion, has been quoted as remarking, "To say that it is tundra today and should be tundra tomorrow when tundra has no economic value doesn't make sense."

Predictably, when the battle began over the TAPS pipeline, the state voiced little concern for the Alaskan wilderness; it displayed only impatience with the delay and was apparently concerned only with sweeping aside the conservationists' objections and getting its own private 800-mile cornucopia built post-haste. Governor Keith Miller called the delay in Senator Jackson's committee "illegal" (Naturally, he had already gone on record as saying, "Nobody is better prepared to protect Alaska's environment than the state of Alaska itself."), and said in August, 1969, "We're making every effort we can to help TAPS get on with the business of building the pipeline. [The delay] has gone on too long already." To speed things along, the state's Department of Natural Resources put out a film that shows, in glorious color, the great pains the oil companies are taking to preserve the Arctic environment. After watching the film's dull camera work, listening to its travelogue narration, and absorbing its schmaltzy background music, all of which evoked all the fourth-rate documentaries one has ever watched, a reporter sitting beside me said, "At least if the oil companies had made it themselves, it would have been a better film." Well, the word in Anchorage was that the oil companies—Arco, to be specific—*had* made it, and the Alaska Department of Natural Resources was only acting as a front.

Whoever made it, the film is an example of the oil industry's efforts to convince a suddenly ecology-conscious public
of its determination to protect the environment. The same
methods are equally suitable for other industries caught un-
comfortably in the spotlight of public scrutiny. The essence of
the film's argument is that various techniques the oil companies would use anyway to maximize their own efficiency and
profit are actually being used to preserve the neighboring
ecosystems. It would require a huge leap of faith to believe,
as the film asks its viewers to do, that the oil companies are
capping gushers and generally trying to keep oil from leaving
the ground uncaptured just because they'd hate to see the
tundra messed up. It's equally hard to believe that the oil companies are using centralized control posts that can run many
drilling operations from a single instrument panel just to avoid
defacing the tundra with extraneous buildings or having to
tear up the tundra by making repeated trips from one drilling
site to another. The contention that these drilling methods are
employed solely to protect the environment is obviously non-
sense. It is true, though, that many steps the oil companies
take to further their own narrow self-interest happen in fact to
protect the environment. "We want a lot of the same things
the conservationists do," says Buddy Morel, the engineer who
is running the TAPS construction project from Anchorage,
"even though we may want them for different reasons."

It has become obvious, though, that a certain amount of
what Walter Hickel once disparagingly called "conservation
for conservation's sake" will have tremendous public-relations
value for the oil compaties and may in fact be necessary if
they are to get permission to build the pipeline now and to do
other things in the future. TAPS and the individual oil companies have commissioned a number of "environmental studies"
and have based their decisions at least partly on the data that

these studies have gathered. Some of the studies, of course, have direct practical value—it is very much in the companies' interest to know all they can about permafrost, or to find out what kind of vegetation can best be grown in the Arctic to protect their work sites from erosion and shield the permafrost from the rays of the sun. But the implication—and it is more than an implication—is that the environmental studies also have the best interests of the flora and fauna at heart. Some of the studies may be scientifically impeccable, but it's hard to escape the conclusion that some are simply conducted for show. Luther Carter reported in the October, 1969, issue of *Science:*

"TAPS is sponsoring a study by the [University of Alaska's] Institute of Arctic Biology on methods of revegetating tundra that has been disturbed by engineering operations. 'They want quick answers, but unfortunately there are none,' Peter Morrison, Director of the Institute, told an interviewer last summer. 'Frankly, I'm not too happy with this kind of research.' " Other TAPS research has been conducted by the University of Alaska's Institute of Arctic Environmental Engineering. On December 7, 1969, the *Seattle Times* carried an interview with the Institute's director, Dr. Charles E. Behlke. "I feel quite strongly that the ecologists and politicians are way overplaying the 'delicate ecology' pitch to either get more research grants or to make political hay," Behlke was quoted as saying. "The fact is that the wealth of a few square miles in the Prudhoe Bay area is quite close to the value of Manhattan Island in New York. I don't feel that the 'delicate ecology' of such a valuable industrial area is so important that it should stand in the way of development.

". . . I do not believe in tearing up the country needlessly, but I do feel that $100 billion to $400 billion worth of oil is far from a 'needless' cause.

"This development means that the great State of Alaska can begin to move forward as a center for industry, scenery, tourism, transportation and government. . . ."

TAPS has also done some environmental research on its own. TAPS project director George Hughes told the Senate Interior Committee in September, 1969, that "We have also engaged the services of Bryan L. Sage, a noted biologist and ecologist, who for several years has been concerned with the biological problems of oil pollution.

"Mr. Sage has organized a multidisciplined ecological survey team to determine the effects of the construction and operating activity on the wildlife population or along the specific pipeline route.

"The team completed its initial survey a few weeks ago and is continuing its effort to determine if other ecology problems exist which should be investigated."

William Houff, a Unitarian minister from Seattle who had spent the summer of 1969 at Big Lake, just south of the Brooks Mountains, told me that he had encountered and spent time with Sage's "multidisciplined ecological survey team." "Any research project has a feel to it," said Houff, who was once a research chemist for Shell, "and this one felt wrong." Houff said that the TAPS team set a gill net, caught a herring, and concluded that herring were all that lived in the lake. The truth is that the lake, which is some eight miles long, also contains trout, pike, grayling, ling cod, and possibly other kinds of fish. (I had no reason to mistrust Houff, but I decided that I couldn't responsibly accept his admittedly partisan reporting as fact. I checked his story out and, after hearing a virtually identical story from an unimpeachable but, unfortunately, unquotable source, I now have no reservations about printing it.)

However suspicious some of the studies and most of the

propaganda may be, the fact seems to be that at least some of the oil companies are now trying to keep their environmental damage to the North Slope to a minimum. Obviously, going in there at all presupposes the view that a certain amount of environmental damage is inevitable and permissible. Beyond that, though, even people who are generally critical of the oil companies' performance on the North Slope have told me that Arco, for one, has been really trying to keep the damage to a minimum. The only explanation I've heard for Arco's relatively great virtue is that Robert Anderson is a sincere conservationist. A couple of people have told me, with wonder, that Anderson has made substantial contributions to Friends of the Earth, the new and rather radical conservation group founded by a former president of the Sierra Club, David Brower. No one has ever satisfactorily explained Anderson's thinking to me. Joseph Fitzgerald, Arco's director of community relations in Anchorage, simply says that Anderson is "ahead of his time."

But even at Arco, not all the top brass is quite so forward-looking as Robert Anderson. When Louis Ream addressed the businessmen in Seattle, he made all the requisite references to conservation, but his voice was heavy with sarcasm. The tone in which he assured his audience of the safety of Arctic caribou left little doubt that caribou are not among the things he stays up late worrying about. When I asked Ream, after his speech, if the main difficulty in getting oil out was posed by the Arctic environment, he said in an irritated tone that "the main problem is the opposition of the conservationists. If environment and technical problems were all, we'd be much further along."

There is no doubt that pressure of public opinion and government stipulations have made the oil companies a lot more conservation-minded than they would have been, left to their own devices and consciences. Take the simple practice of leaving trees to screen the pipeline from the road that's to be built beside it. Says TAPS engineer Buddy Morel, "There's no economic advantage in leaving trees and no economic advantage in not leaving trees. But if the public gets mad at us for not leaving trees and won't buy our gas at the service station, that hurts." Morel says that the government stipulations were really a good idea. He and the people he works with are glad to protect the plants and wildlife in their path, he said, but left to their own devices they wouldn't have done it. They simply had never thought in those terms, and they never would have given much thought to whether or not they were laying their pipe through fish-spawning grounds, for instance, "if someone hadn't pointed out these areas of concern. We *need* people to identify these areas of concern for us," he says, explaining that once the problems have been identified, he and his fellow workers can "go ahead and do it the right way." Morel strikes me as the best possible living advertisement for TAPS: by admitting that, left to their own devices, the oil people probably wouldn't have given much thought to the preservation of Arctic ecosystems, he doesn't strain one's sense of plausibility; and when he says that knowing about the environmental problems involved, he is glad to "do it the right way," he sounds absolutely sincere. I asked Morel if the stipulations and the care TAPS will consequently have to take would set a new pattern for big industrial projects in the United States. "A new pattern has already been set," he said. "We're the first ones who've had to follow it."

8

THE STIPULATIONS that Morel and his colleagues will have to abide by were conceived as the means by which, as Nixon said in a memorandum to Hickel, "We can explore and develop, without destruction and with minimum disturbance, the oil resources of northern Alaska." The stipulations forbid TAPS to pollute the land or water, use persistent or mobile pesticides (including DDT and its relatives), or block the passage of fish or big game animals. Any damage to fish-spawning beds must be repaired. The government can suspend construction in certain areas when wildfowl are nesting, caribou are migrating, or fish are spawning. (Russell Train chuckles at the thought of oil companies' planning their operations around migrations and spawning. It's definitely a thought worth chuckling over.) Any activity that might significantly alter lakes or streams or the quality of their water must be cleared in advance with the Alaska director of the Bureau of Land Management. The pipeline must be built in a way that will "reduce erosive effects" and "prevent degradation of the permafrost in areas where such degradation would result in detrimental erosion or subsidence." TAPS must repair any damage and clean up any oil spills, and it is liable for any damage caused by spilled oil. (It is not financially liable, however, for aesthetic or ecological damage.) The stipulations can be amended at any time. And, as the *pièce de resistance,* if TAPS violates any of the stipulations, the Alaska director of the Bureau of Land Management—and ultimately his boss, the Secretary of the Interior—can "suspend or terminate" construction of the pipeline or operation of the finished line.

A "letter of acknowledgment" from Hickel, which precedes the stipulations, calls them "guidelines that will provide exacting environmental protection during the construction and operation of the Trans-Alaska Pipeline." Hickel has also said, "The stipulations will insure that the wildlife and ecology of the Arctic . . . will be enhanced." ("What, on the little that remains of God's green earth, does he mean by that?" asked the Sierra Club *Bulletin.* "Federally funded Medicare for caribou? How can an 800-mile-long mechanical monstrosity 'enhance' the ecology of the Arctic?")

Train has said that, to his knowledge, "no private construction project has ever been asked to accept such strong constraints or such continuing direct control by the federal government." But a number of people have publicly doubted that the constraints are strong enough to do much good. For one thing, they doubt that the federal government would ever shut down a billion-dollar project (the pipeline is expected to cost $900 million, and the figure will almost certainly rise) just because some subcontractor's employe drives his bulldozer through the wrong part of a stream. Robert B. Weeden, the Alaska representative of the Sierra Club and president of the Alaska Conservation Society, told the Senate Interior Committee in October, 1969, "I am very worried about the system that is proposed in the stipulations for protecting the public's interest. The authorized officer, who, incidentally, would have to be a genius in at least a dozen fields, and each of the six or eight men who serve in his place on the individual construction spreads will have absolute discretion over the process of construction. These men have the authority and responsibility to suspend or terminate construction if stipulations are violated.

"Here you have men embodying and carrying the public interest hundreds of miles from even a regional headquarters, sharing bad terrain and wicked weather and cups of coffee with men they come to befriend. Then someone makes a bad mistake. Question: Will these men put aside friendships, forget the tens of thousands of dollars a single day's delay will cost the contractor, and stop the work?"

Train insists that since the power to shut the pipeline down exists, there's no reason to believe it won't be used—but he concedes that the decision to shut it down would be a difficult one to make. Buddy Morel agrees with Train that the stipulations probably will be enforced, and he says the system is a good one. "Regulations have to have teeth in them," he says, and the power to shut the project down constitutes the sharpest possible teeth. "At first there was talk about setting price tags on streams and so forth and letting companies tear things up as much as they liked, but making them pay for any damages," he says. "I'm against that from two points of view. I'm against it as an oil man because it's less expensive if you tell us what's wrong to begin with and we can just not do it. I'm against it as a private citizen because suppose you set a price tag on a stream of $750,000, but it would cost a million dollars to cross that stream right—well, the oil company would go ahead and *pay* the $750,000. A great deal in this world is determined by economics."

Even assuming that the stipulations can and will be enforced, many conservationists think they provide a flimsy shield for Arctic ecosystems. "The stipulations cover the topics but they don't give the details," says Robert Weeden. The stipulations about erosion control, for example, say only that "the design 'shall include specifications for the construction of erosion-control and drainage features that will minimize the concentration of water and thereby reduce erosive effects.' Then they list a few things the contractors can use. The point is, they don't know if those things will work. You're left hoping the contractors know what they're doing." And, Weeden says, that's not much of a hope. "There's nothing in the stipulations," he says, "that indicates even the people who wrote them understand the problems."

One reason the stipulations didn't include many details is that nobody *knew* many details. "The [Alaska] Department of Fish and Game never used to consider Alaska north of the Arctic Circle as a 'problem area,' because of the absence of industrialization and the large supply of wildlife in relation to demand," Weeden wrote in June, 1969, when he was still a regional research coordinator for the Alaska Department of Game. "Conditions and attitudes have changed since July, 1968," he said, but the oil industry was moving faster than the Department could, and "resource managers scribble plans in the wake of bulldozers." Because the Arctic hadn't been regarded as a "problem area," no one had ever bothered to do much research on it, and, as a result, "biologists . . . know little about game and fish resources in northern Alaska."

There are some 103 major fish streams in the path of the pipeline, says Rupert Andrews, director of sport fish for the Alaska Department of Fish and Game, and no one has ever known for certain what fish live in them. By working frantically, Andrews' department has managed to form a pretty good idea of what lives in the streams, but no one can pinpoint, for instance, the exact times or places of spawning. "We said last summer that we'd need two years to do the job," he says; "not to do a thorough job, just a halfway adequate one.

The oil companies' top management has been very cooperative, but the one thing they won't give an inch on is time."

The Sierra Club *Bulletin* for October–December, 1969, says
that "the single most critical issue of the whole Arctic exploitation question is the fact that so little is known about Arctic
ecology at this time that it is impossible to assess potential
damage. . . . [Secretary Hickel's] department has performed
one of the miracles of our time by preparing a list of stipulations governing the construction and operation of the pipeline
based on ecological information that isn't even known yet. . . .
[Hickel] noted that the stipulations 'are designed to meet *all*
of the environmental and ecological goals set forth by the
department, based on research by its own scientists, independent authorities, and public hearings held in Alaska.' Non-
sense! If the pipeline is allowed at this time it will be installed
on a trial-and-error basis pure and simple. Most of the studies
on just the mechanical phases of pipeline construction in the
Arctic—investigations into the effect of the pipeline on permafrost and vice versa, for example—have not been completed.
Hardly any biological studies have even been started. . . ."

One piece of information that nobody had when the stipulations were being written is contained in a report prepared
by Arthur H. Lachenbruch for the U.S. Geological Survey,
entitled *Some Estimates of the Thermal Effects of a Heated
Pipeline in Permafrost* and generally referred to as the "Lachenbruch report." Lachenbruch, who ran the very scant experimental data available through a computer, predicted:

"A four-foot pipeline buried six feet deep in permafrost and
heated to 80°C. will thaw a cylindrical region twenty to thirty
feet in diameter in a few years in typical permafrost materials.
The principal effect of insulating the pipe," he said "would be
to increase oil temperatures rather than to decrease thawing."

In most places, "thawing will continue throughout the life of the pipeline."

Under certain circumstances, "the entire thawed cylinder would tend to flow like a viscous river and seek a level. . . . The uphill end of the pipe could, in a few years, be lying at the bottom of a slumping trench, tens of feet deep, while at the downhill end, millions of cubic feet of mud (containing the pipe) could be extruded out over the surface. . . .

"Almost imperceptible systematic movements of the thawed material can accelerate the thawing process locally by as much as a thousand times. . . .

"Massive vertical veins of ice . . . are widely distributed in northern Alaska. A pipeline crossing ice-wedge networks at random angles would thaw the wedges quickly and could thereby lose support over considerable spans. A statistical calculation suggests that in typical ice-wedge terrain, conditions which might exceed the design stress of the pipeline could occur on the average of once every mile. . . ."

The Lachenbruch report was so alarming that even after the stipulations had been prepared and signed and the Interior Committees of both the Senate and the House had given their approval, the Interior Department held up the permit for the pipeline—at least, as it turned out, until the oil companies' own research project on the effects of a hot pipeline on permafrost had been completed. There had been questions about permafrost anyway, and the report was more or less the final straw. "We are confident that the state of technology is capable of solving these problems," Hickel wrote to Senator Jackson on December 3, but "at the present time, TAPS has not yet provided us the necessary answers and assurances which would permit us to move ahead."

Whether the Interior Department held up permission on its own initiative or because, as one Washington rumor has it, certain people on the White House staff got alarmed, I don't know. ("Both the Secretary of the Interior and the White House realize," the *Christian Science Monitor* reported on January 12, "that the pipeline is a political time bomb that might explode later should a pipeline break spread oil over the delicate Alaskan tundra.") Permission *was* withheld, and when it didn't materialize by January 1, things around the Interior Department got pretty tense. The oil companies still wanted to start work in March. They had been stockpiling pipe at Valdez since September, and by early December 100 miles of pipe were stacked in the tiny port. And they had received bids for five of the six sections of construction. (Ordering the pipe wasn't an attempt to pressure the government into granting a permit, Arco executive vice-president Louis Ream said on December 12; it was simply a "risk-taking venture by American private enterprise." Ream explained that TAPS had had to order the pipe early because there was a substantial "lead time" between ordering it and having it delivered, and delivery of all 800 miles of pipe would take two years as it was. He also said that the pipe had been ordered from Japan instead of from United States steel firms—Alaska's governor Miller had charged at one point that the steel industry was putting pressure on Senator Jackson to delay approval of lifting the land freeze—because American companies would have needed a full year to tool up for the job, while Japanese factories could start producing it much sooner.) If no permit was forthcoming by the end of January, they'd have to ask for new bids. Permission didn't come, and on January 19, TAPS announced that it was going to ask for new bids when permission was granted.

The TAPS public-relations line is that permafrost is less of a problem than it has been made out to be; it is, in fact, as I was told when I first phoned the TAPS office in Anchorage, a problem created largely by the press. The information in the

Lachenbruch report wasn't current when the report came out, the TAPS people said; the oil companies had considered many of the same facts long before, had since done a great deal of research on the matter, and had worked out solutions to all the major difficulties. Those claims aren't totally implausible. The companies have actually done more research on laying hot pipelines through permafrost than anyone else. Before they decided to build the TAPS pipeline, no one had ever paid much attention to the subject. And Lachenbruch specifically denied any "intention to imply that these problems have not already been considered, or indeed solved, by engineers concerned with them."

But the problems Lachenbruch pointed to were extremely grave—William T. Pecora, director of the U.S. Geological Survey, said in his introduction to the report that "rapid thawing of the permafrost in critical areas could have most significant—and extremely complex—effects on the local environment and upon the security of the pipeline." Lachenbruch predicted, for example, that if the material around the pipe becomes liquid, "it may be difficult to anchor structures to retard the flow [of the liquefied permafrost]. Sheet piling that was installed at regular intervals to dam the flow would have to be anchored in material not subject to liquefaction. If it were not, it might serve only to transfer the hydrostatic-flow forces to the pipe." Also, Lachenbruch said, "Potential problems certainly exist that have not been considered." After Lachenbruch's distressing predictions, TAPS assurances that it had the engineering problems well in hand were received rather skeptically. Shortly after the report came out, William Van Ness, special counsel to the Senate Interior Committee, spoke with astonishment of the oil companies' "sublime confidence" that they would solve all the major engineering problems by February.

Needless to say, TAPS kept trying. TAPS representatives have made much of the fact that permafrost is simply any substance that stays below 32° all year round, not necessarily a substance that will melt if it is heated. (Buddy Morel, once again more credible than other spokesmen, is content with this definition. Some people say that the experts regard permafrost as anything that stays that cold *for two years or more*.) Permafrost can just as easily be gravel or rock as sedimentary soil. If gravel or rock is heated, TAPS people point out, all you'll have will be hot rock or gravel, which won't melt at all. But this is obviously begging the question: the problem that everyone is really talking about has to do with laying a hot pipeline through sedimentary soil with a high ice content— "dirty ice," it's called in the trade. Even in that kind of permafrost, the TAPS people say, a pipeline can be laid safely.

Lachenbruch describes the problem in sedimentary soil as follows:

"The most important difference between permafrost sediments and familiar unfrozen sediments stems from the fact that water generally has considerable strength when it is frozen [ice], and when it is thawed, it does not. When . . . permafrost [which contains more ice than could be absorbed into the soil if it were melted] thaws, it is no longer stable, and the mineral grains tend to settle and to exclude the excess water. . . . Imagine the pipe to be embedded in sediments containing excess ice. As the thawed cylinder grows, excess water will be generated at the advancing interface. If it can migrate to the ground surface as fast as it is generated, the grains can settle down to contact one another, and the thawed sediment will gain increasing strength by consolidation (excluding additional water). If the rate of migration cannot keep pace with the rate of generation, excess water will persist in the thawed cylinder, part or all of which will tend to behave

as a fluid. Predicting which type of behavior will occur is
. . . very complex." TAPS is naturally willing to risk such a
prediction.

Buddy Morel points out, as Lachenbruch does in his report,
that "the rate at which heat is applied to the soil is important.
If a 170° pipe were put directly into the soil, there *would* be a
large run-off, but the pipe would heat up slowly, and water
should escape without damage." Morel claims that there's no
place on the entire route where the permafrost can't safely support the pipe, either buried or raised on pilings. Refrigerated
pilings are a possibility in extreme cases, but there are also
several kinds of unrefrigerated pilings that use heat convection
to literally freeze any ground they're placed in. According to
Morel, the permafrost in the Copper River Basin, often considered the trickiest on the route, has turned out, on further
examination, to be pretty safe: the soil there is relatively dry,
so that even if all the ice in the permafrost melted, the soil
could absorb it easily, and the result wouldn't be a liquid mass
but simply unfrozen ground. (Other interested parties were
less convinced than TAPS about the safety of the Copper
River Basin. At roughly the same time Buddy Morel told me
that, Senator Jackson said that the ice content of the permafrost in the Copper River area varied so frequently and so
abruptly that even sample holes drilled a few feet apart wouldn't
accurately indicate the ice content of the soil in between.)

Permafrost isn't the only potential hazard along the route,
but Morel says that even if an earthquake hit the pipeline
—two thirds of the pipeline route runs through a prime
earthquake zone—the pipe itself is so sturdy and so flexible
that it would probably "crimp like a soda straw before it
broke. The earthquake of 1964 wouldn't have broken it," he
says. Still, he concedes, "A break *is* possible. I'd be a fool if I
said it wasn't." But he thinks the chances that a break will

occur are minuscule. And, he points out, TAPS doesn't want
the pipeline to break any more than the government does or
the conservationists do. "If you build a $900-million project,"
he says, "you want to make sure it's going to stand."

The oil companies probably will make sure. Chances are
that the pipeline won't disappear into the permafrost or break
in two—which is not to say that no oil will escape, or even
that the oil company higher-ups *think* no oil will escape. Arco
vice-president Louis Ream straightforwardly told the businessmen in Seattle, "There *will* be spills." But probably no spill
will be catastrophic. The real questions aren't whether or not
the pipeline will collapse entirely but: (1) whether even a
slight chance of a major spill is worth taking; (2) whether the
effects of constructing and operating even a sound pipeline
should be tolerated; and (3) whether the object of building
the pipeline—removing and marketing North Slope oil—is
something to be sought or avoided.

There are those who think the public should figure the odds
of a major spill a bit differently from the oil companies. "The
oil industry is betting that in the time required to pay off the
cost of the pipeline there won't be any major breaks," Robert
Weeden says. "TAPS has weighed the risks of pipeline break-
age against profits and it has decided to take the risk," he told
the Senate Interior Committee. "It is saying in essence that
over the economic life of the project, which means this pipeline will have paid for itself in a matter of eight or ten years,
they feel there is not enough risk of severe damage . . . to
warrant them stopping it at this time. I am not convinced that
the public, which lacks the lure of profits, can or should accept
[the] same hazard." Weeden has suggested that "if there *is* a
major spill in the next twenty years, the economic loss may be
negligible," but the ecological damage will almost certainly be
immense. He says that in their decision to build the pipeline,

the oil companies have used the same reasoning the Atomic Energy Commission employed when it decided to permit underground nuclear testing on Alaska's Amchitka Island, right over a major earthquake fault. The reasoning, he says, boils down to: "We've gotten away with it so far, and we're betting we can get away with it a little longer."

What damage will even a sound pipeline do? Well, for one thing, to put it crudely, a virgin wilderness that has an 800-mile pipeline rammed up it is no longer exactly virgin. And it won't just be a pipeline, either. A construction road will parallel the pipeline from Livengood, just north of Fairbanks, to the North Slope. The fifty-three-mile section between Livengood and the Yukon River has already been started. The TAPS road will open great untouched areas of interior Alaska to campers, with their attendant litter and fires, and to small-scale oil and mineral exploration companies that literally can't afford to take many pains with the environment.

(The TAPS road isn't the first, but its predecessor was built too badly to have been of much use. In late 1968, Walter Hickel, at the end of his term as governor, appropriated funds for a highway from Livengood to the Slope. It was to be a winter highway, for use when ice made it impossible to transport materials by barge. There is little evidence that anyone but the Alaskan trucking industry wanted the road, and there is considerable doubt in Alaska that the appropriation, which Hickel accomplished when the legislature was out of session, was entirely legal. The road was completed by Hickel's successor as governor, Keith Miller, and named the Walter J. Hickel Highway. Some of Miller's critics say that naming the highway for Hickel was the only smart thing the governor has ever done. Because the highway itself was a disaster. The only feasible way to build a winter highway in that terrain is to make the roadbed by piling up snow. Instead, the state made its roadway by bulldozing a trench. The permafrost at the bottom melted, trucks were stuck for days at a time, and the whole grotesque thing was in use for only six weeks. The highway has since been revived, at considerable cost to the state, and is now used by trucks going to the North Slope. Heavy equipment is shipped over it, but almost no other supplies. The ride is rough, and the cost is prohibitive.)

William Van Ness says that the road will "open the interior to every 'gypo' driller who wants to prospect on federal land. If you know how, big corporations are easy to regulate," Van Ness says, "but 'gypos' *can't* be regulated. They can't afford the extra expense of conservation. The only way to save interior Alaska from them is new legislation."

A TAPS man with whom I spoke agrees. "There was a great hue and cry at first about letting the 'little guy' into the North Slope fields," he says. "People complained that only the big companies could afford it. Well, the 'little guy' will *really* mess up the country. He operates on a shoestring, and he can't afford to paint oil tanks or clean things up. In East Texas, the Arco tanks are all painted, and they're beautiful—or at least as beautiful as something like that can be, compared to nature. But if you look over onto the next lease, which is held by an independent operator, it's a *mess.*"

It's by no means certain that many small mineral exploration companies will take advantage of the TAPS road in the immediate future. Governor Miller told me in January, 1970, that he certainly *hoped* the road would open parts of interior Alaska to mineral exploration, but both Robert Weeden and Robert Atwood, men who don't agree on many things, pointed out to me that Alaska already *has* roads to some remote areas that are rich in minerals, and that nobody uses them. The danger of a real mineral rush may lie many years ahead. Any exploration damage done sooner may be the work of a relatively few people.

But mineral exploration isn't the only incidental use that has been forecast for the road. The Interior Department said in November, 1969, that once the road is built, "there will be increases in demand for townsites, for residential and industrial purposes, rights-of-way for transportation and utility needs, airfields and recreation developments."

Some conservationists believe that even before it is ready for use, the road may have serious effects on the environment. On March 26, 1970, three conservationist groups—the Wilderness Society, Friends of the Earth, and the Environmental Defense Fund—applied for an injunction to prevent Walter Hickel from granting a permit for the pipeline as then planned. The conservationists' brief said, "The construction of the pipeline road will require 12,329,000 cubic yards of gravel, and other pipeline appurtenances will require an additional 1,000,000 cubic yards of gravel. The gravel for this purpose will be taken from the rivers, streams, and lands of the public domain at various points along the route of the pipeline. The mining of this gravel from stream beds and terrestrial sites will cause additional erosion and its attendant problems. It could destroy areas of important fish-breeding habitat and riparian wintering areas for moose, other mammals, and birds." In general, the brief concluded, "The pipeline will have a substantial adverse environmental impact on a significant portion of the Alaska wilderness. The impact of the pipeline and the pipeline road will leave ugly scars and exposed pipe and facilities throughout the area and will destroy the wilderness character of the area, subjecting it to the pressures of exploitation and development of mechanized civilization."

The pipeline and its road aren't the only contact Arctic

Alaska will have with the oil industry. The Prudhoe Bay area will certainly spawn a huge industrial tangle of oil rigs and pipes. The oil companies are already pressing the federal government to let them start exploring for oil in the Arctic National Wildlife Range, east of Prudhoe Bay, between the oil fields and the Canadian border. A well-informed source in the Interior Department told me in January, 1970, that within two to four years, the pressure would be tremendous. (Walter Hickel had already decided, when he was governor, that it would be a good idea for the oil companies to go into the northwestern part of the Range, the part closest to Prudhoe Bay, just to see how much oil was there.) The Canadians want a pipeline that would go east from Prudhoe Bay to the Mackenzie River, then down through Edmonton to the midwestern United States, and that would just incidentally be open to oil from Canada's own northern reserves. Such a pipeline would have to go either through or just offshore of the wildlife range. There will certainly be extensive exploration for oil and possibly for minerals in other parts of the Alaskan Arctic, which has huge deposits of both.

The oil companies began exploring in the Beaufort Sea, the part of the Arctic Ocean that lies above the North Slope, in the summer of 1969. The exploration involved underwater blasting and other major disruptions, so the Alaska Department of Fish and Game, which was already hard-pressed trying to catalogue the wildlife in the path of the TAPS pipeline, sent two biologists up to see what lived in that part of the Beaufort Sea. The biologists and their boat vanished in a bad storm, taking many of their research findings with them, but not before they had reported that the area contained a commercial-size run of chum salmon that nobody had known about before. At this writing, the state of Alaska plans to lease some 100,000 acres of the Beaufort Sea for underwater drilling in July, 1970, so that the salmon and the other underwater wildlife there may be in for a bit of a shock. The state also plans to lease some 70,000 acres in the more southerly Bristol Bay, which contains the largest red-salmon fisheries in the world and the greatest seasonal concentration of waterfowl in North America. Robert Weeden, who is very concerned about Bristol Bay, also points out that the huge tankers carrying oil south from Valdez, will seriously endanger the sea otters and other wildlife in Prince William Sound. "We know that for every thousand gallons of crude oil transported across the ocean, one gallon has been spilled by tanker accidents," he says; since plans call for a million gallons a day to be shipped from Valdez, spills of 1,000 gallons a day may result. And the oil won't be spilled a gallon at a time, he says. If it escapes, it will escape in huge quantities.

Despite the misgivings of Weeden and other conservationists, the really big development after the North Slope will almost certainly be offshore. Offshore drilling in the rough Arctic waters will inevitably carry great risks of Santa Barbara-style spills. And, as a few people morbidly insist on pointing out, a spill in the Arctic Ocean could spread dark oil over many miles of ice, the oil could then soak up solar heat that is usually reflected from the white ice, and the resulting increase in temperature could melt the Polar Ice Cap.

9

OIL ISN'T THE ONLY INDUSTRY that is causing some conservation-minded Alaskans to lose sleep. The fishing industry, which has long been important there, has fished itself almost out of business. The logging and pulp industries are doing their share to plunder the wilderness; in fact, state conservation officials worry more about them than about oil. The Army Corps of Engineers is eager to do its part, too: the Corps-proposed Rampart Dam would flood a huge undeveloped area in the eastern part of the state. But the pipeline has received far and away the most national attention.

The two objections I've heard raised most often are, first, that the Arctic's ecological balance is so delicate that it should not be tampered with without a great deal of study, and, second, that the pipe will prevent the migration of caribou.

The caribou argument is spectacular—400,000 large animals trapped behind an 800-mile-long, six-foot-high metal wall makes quite an image—and when there was talk of laying the pipeline above ground it made sense. But now that TAPS plans to bury all but short sections of the pipe, the caribou will probably be able to migrate to their hearts' content. Even Robert Weeden and the Eskimos of Anaktuvuk

161

Pass doubt that the pipeline will harm the herds. The caribou will almost certainly make out all right.

As for Alaskan ecosystems in general, the pipeline and its feeder lines, roads, and pumping stations will all certainly tear hell out of anything directly in their path, but Alaska *is* huge, and the pipeline system won't singlehandedly turn the state into a desert. (Of course, if one regards the pipeline as a first step, it's pretty formidable.)

For the people who oppose the pipeline most bitterly, both arguments are probably peripheral, really just attempts to substantiate a much vaguer but more basic feeling that comes down to "we've already ruined the rest of the country—why do we have to start messing up Alaska, too?" Russell Train concedes that this is a valid feeling, and, he says, more or less because of it, the government has paid special attention to protecting wilderness areas along the route.

I believe that Train is sincerely concerned, but that's hardly the point. Did his task force ever seriously consider the possibility of not building the pipeline at all? Did it seriously weigh the possible benefits of the pipeline against the possible costs? There's no evidence that it did.

After the stipulations were written but before the Senate Interior Committee gave its approval, Senator Jackson sent the Interior Department a list of questions. "What public—as opposed to private—benefits does the Department see in construction of the pipeline at this time?" he asked. "What public risks does the Department see? Is it the Department's judgment that the benefits outweigh the risks and therefore justify going forward at this time?"

To the last question the Interior Department replied, "The Department's communication with and testimony before the appropriate committees of Congress concerned the lifting of the land freeze for the pipeline right-of-way, and not the granting of the actual permit." In other words, "We won't tell you."

To the second question the Department replied, "The major public risk is a potential oil spill." Then, asked about "the magnitude of the worst possible spill," the Department answered, "The capacity of the pipeline is reported to be 500,000 gallons per mile. The size of a possible spill would depend upon, among other factors, the extent of the break, velocity of oil movement, reaction time for shut-off, viscosity of the oil, and amount of oil that could not be contained." In other words, no one has ever bothered to figure out how much oil might be spilled if the pipeline broke. The government hasn't and the oil companies haven't.

When pipeline project director George Hughes appeared before the Senate Interior Committee, Jackson asked him, "If a break occurs, what is the fastest that you could cut it off, and what would be the maximum amount of oil that would spill?"

Hughes replied, "After the system would be cut off it would depend entirely upon the location of the break . . . and the topography of the terrain. . . . Naturally, if the break were to occur at the top of the hills you would lose no oil after it was shut off but if it occurred in the bottom you would naturally drain the section of the line feeding to this. Of course the size of the break enters into it, too."

Jackson then asked, "The distances between the automatic shut-off stations are quite substantial, are they not?"

"Yes," Hughes said. "Well, we . . . will have five pump stations initially [for the 800 miles of pipeline] and twelve at the completion of the system. However, it is the terrain that determines the drainage and not the distance between the pumping stations."

"Yes," Jackson said, "but let us take the worst situation. You always prepare for the worst, you know. How much oil could spill before the automatic cut-off would be effective? One of the reasons I raise this question, of course, is so that we know exactly what we are getting involved in."

"Well," Hughes answered, "if we were to have a major break, the automatic cut-off should cut off instantaneously. These are electronic devices and they operate in microseconds, and it would be just a matter of the time of closure of the valve. These are large valves. They are thirty-some feet high, they weigh 60,000 pounds, and obviously you can't close one of these valves instantaneously. So there is a several-minute period of closing the valve until you get it completely closed off."

"What I am getting at," Jackson persisted, "is that you would have an area that is not covered by the automatic shut-off, so that there would be quite an outflow. . . . The Department of the Interior has indicated in their letter to us that it would be an outpouring, I think, . . . of a half-million gallons per mile."

"The pipeline will hold about that quantity of oil per mile," Hughes said, "and if you drain a mile then you would naturally spill that much, but————"

"But these automatic shut-offs leave a substantial area that would not be covered," Jackson insisted, "and, depending on the terrain, you would have an outflow there that could cover several miles."

"That would depend on the break," Hughes said.

Strange that no one could calculate just how great the greatest inherent danger in the pipeline was, that nobody bothered to figure out and consider the dimensions of what could be the biggest single ecological disaster of the decade.

The Interior Department had no such trouble computing the possible *benefits* of the pipeline. When Jackson asked what those benefits were, the Department replied: "Consumption of petroleum on the West Coast, including Alaska and Hawaii, has been and is continuing to grow rapidly. West Coast demand for oil has grown from 1,500,000 barrels per day in 1965 to about 1,900,000 this year. By 1972, West Coast consumption of oil is expected to increase to about 2,200,000 barrels per day. [It's evidently easier to calculate how much petroleum the entire West Coast will use two or three years hence than to calculate how much oil can spill from a busted pipe.] Current production of oil on the West Coast, including Alaska, is about 1,300,000 barrels per day—and it now appears that, except for northern Alaska, the production will not increase between now and 1972. A delay in [constructing] the pipeline will aggravate the gap between West Coast consumption and supplies of oil."

In other words, North Slope oil is needed to feed the Los Angeles smog. Feeding the Los Angeles smog is unequivocally considered a "public benefit." And the public will benefit even more if the smog is fed with oil produced right on the West Coast of the United States.

At this point in history, when automobiles are choking America's cities and smog is choking their inhabitants, the Interior Department's conception of "public benefit" may seem slightly bizarre; it is at any rate open to serious question. But there is really no point in questioning it; any discussion of the pros and cons of the pipeline will be purely academic unless there is some chance that construction of the pipeline will be forbidden. There never has been such a chance. The relevant government officials have lacked both the legal machinery and the will to unconditionally forbid construction.

Senator Jackson says that to block construction of the pipeline, Congress would have had to pass special legislation. Besides, although he doesn't think there's any justification for taking the oil out *now*, he does think that we need the oil eventually. He hopes that it will lessen our dependency on the oil-producing Arab states (he thinks that the coup in Libya makes a domestic source of oil more important than ever). Although he doesn't think it will provide an alternative to Middle Eastern oil for daily use, in an emergency it will be able to supply us and the Japanese and, if sending ice-breaking tankers through the Northwest Passage proves feasible, Western Europe as well.

Russell Train, too, says there was never any chance of the Senate's deciding that the pipeline simply shouldn't be built. When Train testified before the Senate Interior Committee, Gaylord Nelson of Wisconsin asked him, "Has anybody raised the question of why we ought to exploit this resource at this time?"

Train replied: "Well, I certainly have heard the question asked, Senator, yes. If you were to ask me whether the Department has made a decision that the oil should come out, and I think that is implicit in what you have said—first, it is not a decision that is entirely up to the Department of the Interior.

"Certain commitments have been made by the private sector in terms of development. I think something in the neighborhood of some $300 million of exploratory development activity has already occurred on the North Slope. The companies have recently paid something in the neighborhood of $900 million for additional lease interests.

"I believe that there has already been an investment in the neighborhood of perhaps as much as $200 million in connection with the pipeline and related matters, so that in a sense the private sector, at least, has made a decision that this is an important resource which it expects to develop, and this has been a traditional way in which such decisions have been made in this country. . . .

"There is no question that the time frame within which we

all find ourselves in this problem [the Interior Department was working to get the pipeline approved quickly enough to let the oil companies start construction work in March '70] has been created in substantial part by the timing of the companies' own investments and decisions.

"I think that whether we would have moved in exactly the same way and created a similar time frame is very much open to question, but that really isn't before us. We are confronted with a situation which exists."

Train says that because the decisions about whether and how a given resource is to be exploited have been made exclusively by private industry in the past, the government stipulations attached to the pipeline—which don't touch the "whether" but at least constitute a slight modification of the "how"—represent a historic landmark. But, he adds, there is no legal machinery set up that enables the government to make the *basic* decisions of whether and how. In the case of the pipeline, the oil companies in effect put an application for a permit on the Interior Department's desk, and that application is all the Department can consider. Besides, he says, "The people who simply say 'don't take the oil out' provide no alternative. There's no substitute for petroleum. Our transportation system needs it."

So what do we have? Russell Train, who was brought into the government as the Nixon Administration's house conservationist, who is now chairman of the Environmental Quality Council, who was chairman of the task force that produced the unprecedented stipulations for the building of the TAPS pipeline, says unequivocally that there was never a chance that the line wouldn't be built, that in fact the government had no way of preventing it from being built. And Senator Henry Jackson, chairman of the Senate Interior Committee, the man who introduced the Environmental Quality Act, agrees. You'd prob-

ably need a lot more than Diogenes' lantern to find anyone
who disagreed. "There's no power on earth that could have
kept that oil from coming out," says former Attorney General
Ramsey Clark, who, as an attorney for the Alaska Federation
of Natives, has kept an eye on the North Slope and, as a
former Assistant Attorney General for Public Lands, knows
how such things work.

Maybe we do need North Slope oil and the TAPS pipeline
as soon as we can get them for the sake of our transportation
system or our military security or the well-being of our industry. But maybe the price, considered as part of a complex
system of development, is too high. The people who introduced and support ecology as a political issue have one essential message: "If we screw around with nature much more, we
will become extinct." Measured against the prospect of extinction, mechanized transportation, national defense, corporate
profit, and many other old favorite justifications for the exploitation of land and resources seem pretty trivial. But no
one can think seriously of weighing the value of a wilderness
against the value of a pipeline, or the potential damage of oil
spills against the potential corporate profit that lies in the per-
petuation of smog—unless there *is* some power that can, and
is willing to, "keep the oil from coming out." If President
Nixon, or any other politician who has jumped on the ecology
bandwagon, really cares about the environment, he will try to
set up such a power. If he doesn't try, if he simply attempts to
cash in on the issue by spending $10 billion of the taxpayers'
money to clean up polluted water (what could take less guts
than cleaning up dirty water?), then he will have clearly
shown that his real concern is political profiteering, and his
professed concern for the environment is simply bullshit.

10

Today, June 16, 1970, when the application for the TAPS pipeline has been pending for exactly one year, this is how things stand:

Native land claims seem very close to solution. The Senate Interior Committee finished work on a claims-settlement bill on May 13, and there seems to be a serious chance that the bill will make its way through Congress before Labor Day. The committee's April proposal has been amended to retain health services and educational funds now administered by the Bureau of Indian Affairs after the BIA's operations in Alaska are phased out, to give a regional corporation surface rights to an additional 500,000 acres, and, to give title to two additional townships around Barrow to the natives of the North Slope. The extra surface rights mean very little since they don't include mineral rights and the natives use the surface anyway, but the two extra townships, which lie within Pet Four, may include valuable reserves of oil. The natives are hopeful that the bill will pass through the Senate and House virtually unchanged. (Late note: On July 15, the Senate passed the land-claims bill, which had been amended to give the North Slope Eskimos full ownership of their extra 500,000 acres. The Arctic Slope Native Association had announced

two weeks before that its attorneys would try to kill the bill
unless the amendment was added. The announcement hadn't
exactly made Congress quake, but Alaska's Senators Stevens
and Gravel hadn't wanted another delay, and after they had
been waylaid and harangued by the Eskimo leaders during a
visit to Barrow, they had agreed to introduce the amendment
jointly. Senator Jackson gave his support, the amendment was
introduced, and the Senate passed it in 30 seconds.)

The pipeline project, and therefore the whole schedule of
North Slope oil development, is as much in the air as ever.
It seems certain that completion of the pipeline has already
been delayed at least a year. Two preliminary injunctions still
bar Hickel from issuing a permit, and Hickel says he isn't
ready to issue one anyway. The injunction granted on behalf
of the Indian village of Stevens states that "the consent of the
proper tribe or band officials has not been obtained" for a
right-of-way through some of the land claimed by Stevens Village, and that if any right-of-way is granted through that land,
"the members of the tribe or band who it now appears owns
such land . . . would suffer irreparable injury." The second
injunction was granted to three conservationist groups on two
grounds: first, that the Interior Department hasn't complied
with the National Environmental Policy Act, which requires
a complete assessment of the ecological impact of any government or government-endorsed project that will significantly
affect the environment; second, that TAPS has asked for a
100-foot right-of-way for the pipeline, but the Mineral Leasing
Act of 1920 limits pipeline rights-of-way through federal land
to fifty-four feet.

The Interior Department has given additional reasons for
withholding permission; it claims that TAPS is underestimat-
ing the effects of a buried pipeline on permafrost and that
much of the line should be built above ground. Also, TAPS

still hasn't come up with a full blueprint for the project, and Walter Hickel insists that a "thorough engineering and design analysis" is needed. According to Lewis Lapham, in the June, 1970, issue of *Harper's,* the president of the Alaskan Senate, Brad Phillips, said in January, 1970, "that he'd spoken that afternoon to Secretary Hickel in Washington. He reported that Hickel was as anxious as anybody else in Alaska to begin work on the pipeline, but [TAPS] hadn't yet supplied the Department of the Interior with the necessary papers. 'He needs maps and the geography,' Phillips said. 'He doesn't care if they draw it on toilet paper, but they've got to give him something specific.' "

Probably no one has been more disturbed by the delay in pipeline construction than those good citizens of Alaska who hope to profit mightily from both the oil and the construction project itself. Accordingly, the state government has tried to get around Hickel's objections and both injunctions. The state attorney general's office sent a man to Stevens to try to talk the villagers into dropping their lawsuit. He didn't succeed. Governor Miller announced on April 5 that if TAPS couldn't build the pipeline road, the state of Alaska could, and he was willing to do it provided that the oil companies would reimburse the state when the road was done. Walter Hickel said that as he understood the law, the state probably *could* do it. The oil companies tentatively approved, but they were reluctant to commit themselves to a road before the pipeline was approved and its route finally decided.

The state made its grandest effort in May. Governor Miller used $30,000 from the state's special fund to charter an airplane and send a delegation of 120 Alaskan political, business, and labor leaders to Washington, D.C., to ask politicians and industrialists to get the pipeline moving. The day before the delegation left, the attorneys from TAPS were in town, and

Governor Miller was too busy meeting with them to receive Ramsey Clark, who had come to Juneau for a meeting of the Alaska Federation of Natives. Hickel, whose political standing in Alaska had been seriously damaged by the delay, rode along on the plane with the Alaskan delegation.

(The Alaskan delegation was joined in Washington by fifteen representatives of the Seattle Chamber of Commerce. Seattle businessmen have felt all along that they have a particularly big stake in North Slope development. Seattle will be important as a staging and shipping point for supplies sent to the North Slope, and a great many business people in the area hope that it will also become the terminus of a pipeline east to Chicago and New York. No one seems very sure just how being a pipeline terminus would benefit the area—there are visions of petro-chemical industries, etc., but nothing very concrete. Nevertheless, an awful lot of business people seem to think that *some* good would surely come of it and that, therefore, the eastward pipeline is very much to be desired. On the other side, Brock Evans, the Northwestern representative of the Sierra Club, has said that a pipeline will be built east from Puget Sound "over [his] dead body" and that he'll do anything he can to keep Puget Sound from becoming another northern New Jersey. If the oil companies decide they do want to build that eastward pipeline, there may be quite a battle.)

The Alaskans had an appointment to talk with President Nixon, but Nixon cancelled it at the last minute because he was busy with the invasion of Cambodia. They did see Senator Jackson, who reportedly told them he didn't favor the kind of environmental protection that simply stopped development; and they also met with William T. Pecora, head of the U.S. Geological Survey, who told them that as things then stood, he thought the pipeline had only a 50-per-cent chance of holding up. "Hell," *The New York Times* quoted him as saying, "I wouldn't build anything with that chance." But Pecora went along with Hickel, who had announced at the University of Alaska on Earth Day that "the question isn't whether but how the pipeline will be built, and the Interior Department's object is simply to make sure that the line can be built safely." When the delegation returned to Alaska, Governor Miller promptly asked the state legislature for $120 million to build the TAPS road. The legislature voted, right before it adjourned, to spend the money, but only on condition that TAPS would refund it within five years with 7½-per cent interest, whether or not the pipeline itself was ever built. TAPS refused those terms. Miller then called an emergency session of the legislature to figure out some other way to speed the construction project along.

Alaska is really in a bind; the state has been counting'on getting oil money and construction money right on schedule, and it is Alaskans more than anyone else who have been left hanging by the delay in starting construction. The week before the delegation left for Washington, the Anchorage *Times* headlined a story from Juneau: "State Budget Linked to TAPS Timetable." The story said, "The chairman of the House Finance Committee, now reviewing the Special Monetary Committee's package of bills, said that 'a major policy decision' must be made before the legislature can determine what should be done with the $900 million oil-sale bonus money.

"Rep. Bill Ray, D-Juneau, said the decision now hinges on the question of whether construction of the Trans-Alaska Pipeline will be delayed for one year or even longer.

"A delay in construction of the pipeline would mean a delay in projected oil revenues from the North Slope.

"Ray said the legislature has to know what funds might be needed to finance the affairs of the state in the next few years before committing funds 'to a tomato can in the back yard.' He referred to 'locking up' $500 million in a permanent fund, as proposed by the governor, or the $200 million housing-loan investment fund proposed by the House Monetary Committee."

The *Times* also reported that "Potential pipeline haul-road contractors are generally cutting back to skeleton maintenance crews as they await word to begin construction of the haul road from the Yukon River to Prudhoe Bay." The result, the article said, "could be an unemployment problem of major proportions in Fairbanks. The State Department of Labor estimated more than 1,500 unemployed registered at Fairbanks now. [The Department says] the figure could jump to 3,500 or more by next month unless the road goes."

Having put all its eggs in one basket, and having had delivery of the basket held up indefinitely, the state is understandably impatient.

The oil companies themselves are probably in less of a bind. Not having mesmerized themselves with civic-boosting optimism, they were certainly not shocked when March 1 came and went without the issuance of a permit. On December 12, 1969, Seattle transportation expert Fred Tolan told the businessmen's conference in Seattle that "Most oil companies feel there is a serious question that the pipeline can be completed and usable in 1972. Most of them feel that even under optimum conditions they cannot depend on a usable facility before 1973." Earlier, Tolan had said that although "a hoped-for construction starting date is March 1, 1970 . . . the possibility of delay is very real, for it is almost inevitable that there will be one or two court attacks against an order allowing construction of the pipeline. For that reason, it is anticipated that 1970 will see the start of the pipeline but the biggest efforts [by the oil companies] and the biggest movements [of supplies to Alaska] will be in 1971 and 1972."

Some of the oil companies involved on the North Slope may even have been grateful for the delay. After the scheduled date for starting construction of the pipeline had passed, *Oil & Gas Journal* reported that there was a lot less drilling on the North Slope than there would have been if the pipeline had been approved and that some companies might welcome the relatively slack period as an opportunity to build up the capital reserves they had depleted in the September, 1969, lease sale.

One company is certainly less than distraught. It has been accepted for a long time that Standard Oil of New Jersey, the parent company of Humble and the largest oil company in the United States, is in no real hurry to see North Slope oil hit the market—that, in fact, Jersey Standard would like to see the oil come out gradually and not in quantities large enough to depress the market for oil from Jersey Standard's numerous other sources. It is a little easier to understand how the government has withstood pressure from the concerted forces of the oil industry to get the pipeline started if one remembers that the biggest oil company of them all is in no particular rush and can't have been pushing very hard.

The company in the biggest rush would seem to be Atlantic Richfield. Arco certainly isn't happy about the delay, but that's not to say that the company is in any imminent danger of losing its corporate shirt. The day after federal Judge James Hart granted an injunction to the three conservationist groups, Robert O. Anderson was asked at a press conference if there was a point of diminishing returns, a point in time beyond which it simply wouldn't pay Arco to go ahead with the pipeline. Anderson said that there was no such point, and Arco had no intention of abandoning its plans. "We think we have a property there of considerable economic value," he modestly explained, a property that is furthermore "a national asset." He assured his questioner, "We'll get the oil out someday."

It would certainly have been bad tactics for Anderson to say he was worried—any expression of concern would have encouraged his opponents to fight harder than ever to delay construction—but there's no reason to believe that he was just putting up a brave front. The April, 1969, issue of *Fortune* carried a quote that goes a long way toward explaining his lack of alarm: " 'People rant and wave their arms about the billions needed to get this thing going,' says the executive vice president of one of the major companies working the Slope, 'but it's nothing if the oil is there. We used to say 100,000 barrels a day hides all the mistakes, and if you can get one million barrels a day in Alaska, nothing you could do could hurt this thing."

11

IT MAY HAVE BEEN inevitable that the conjunction of Alaska, which has been idealized in American fiction and imagination as the Frozen North of Eskimos and sourdoughs and has been too remote for reality to intrude on the image, and the oil industry, which has been idealized in its own way as the symbol of fantastic wealth rapidly acquired (oil-industry spokesmen are always trying to convince people that this isn't so, but they're never very successful), should produce easily idealized conflicts. The land-claims issue has often been interpreted as a conflict between innocent, fur-clad Eskimos peacefully harpooning seals in a State of Nature and fast-talking men with Cadillacs and big cigars who want to kick the Eskimos off the land, then make millions of dollars from oil wells drilled among the abandoned igloos.

(Many white Alaskans have naturally found it hard to see the conflict in these symbolic terms. For one thing, white Alaskans have been closer to the natives, have known or at least talked with them, have seen them dressed unromantically in ragged work clothes or conventional business suits, have lived in a society in which the "shiftless" native and the drunken native are standard stereotypes. And white Alaskans don't see the men with the big cigars as the natives' only opposition, either. Rightly or wrongly, they feel that they themselves have a stake in the matter, that they themselves have something to lose. They have a proprietary feeling toward the land over which they've hunted and fished, and they have a proprietary feeling toward the wealth that North Slope oil is expected to bring into the state; many white Alaskans seem to feel this wealth is theirs by divine right. They have felt that the natives' demands threaten their own interests—just as Congressmen from states with big Indian populations have felt that a generous settlement of Alaskan land claims would threaten *their* interests—and there is nothing like vested interest to bring that kind of symbolic conflict down to earth.)

The pipeline controversy has been interpreted as basically an attempt by the same fast-talking men with Cadillacs and big cigars to destroy the forest primeval. To both sides in both conflicts, the oil development on the North Slope has been a symbol—as well as an important part—of a whole development process. This process can be seen either positively, as man's conquest of nature, his construction and maintenance of rapid transportation systems and efficient machines, his amassing of national, corporate, and individual wealth; or negatively, as the process that has already destroyed most of the wilderness and aboriginal cultures on this continent, produced unbreathable air and undrinkable water, sliced and strangled the cities with freeways, and brought a material prosperity to some that makes no one happy.

People who take the positive view naturally want to aid the process by removing all impediments to it. The adoption of environmental safeguards for the pipeline and the settlement of native land claims are essentially efforts to remove obstacles, so that the process can proceed with minimal legal and political friction. The land-claims issue has proved the more amenable to this approach. The final question in the land-claims issue was whether the natives would be able to join the process as active participants or would be overcome by it as passive victims. The settlement proposed by the Senate Interior Committee is clearly an attempt to further the former possibility. The natives at least have the option of joining; the wilderness does not. If the wilderness is made part of the development process, it will cease to exist.

Perhaps the wilderness has received more public support than the natives because many people in the United States who view the development process negatively have trouble identifying with the natives but feel themselves in much the same position as the land—they feel that, in significant ways, they can't become part of the process and yet can't escape its effects. For them, the land that is threatened by the pipeline is a logical rallying point. They aren't saying just "save the wilderness." They're also saying "stop the process."

It's most unlikely that these people and the oilmen and politicians who are busily devising environmental safeguards so that TAPS can get on with its work will ever have a meeting of the minds. It would be unrealistic and perhaps unfair to expect even the most enlightened of oilmen—and probably, with a very few exceptions, the most enlightened of government officials—to make the philosophical leap necessary to consider seriously the notion that the process should be stopped. They have profited from it already; they expect to profit from it in the future. So the enlightened oil companies will continue to fire people who throw candy wrappers out of pick-up trucks on the North Slope, and to believe that if all interested parties just sit down and reason together, the environmental problems will be solved. And enlightened public officials will go on drawing up stipulations. And a small but appreciable segment of the American people will go on thinking uncharitably that the oilmen and officials are greedy fakers, or, somewhat more charitably, that they have simply missed the point.

Index

Aboriginal use and occupancy, 101, 102, 105, 106, 119, 124, 126; *see also* Native land claims

Ahgook, Jack, 88, 94

Alaska, 20, 48, 67, 73, 146, 149, 169
  described, 43, 45, 47, 51, 56, 177
  economy of, 21, 43, 46, 51–54, 71, 72, 99, 111, 137, 140
  government of, 11, 13, 51, 57–59, 61, 75–77, 80, 83, 90–94, 112, 115, 117, 118, 123, 125, 126, 136, 148, 156, 159, 160, 171, 173, 174
  history of, 6, 43, 55, 61, 62, 65, 75, 76, 106
  land in, 31, 51, 110, 123, 128, 129, 138, 150, 156–59, 161, 162
  natives of, 21, 61, 62, 69, 70, 74, 77, 78, 80, 81, 85, 86, 95, 100, 101, 106–8, 111, 112, 115, 116, 119, 125, 126
  officials in, 11, 57, 75, 79–81, 89, 91–93, 108, 115, 117, 123, 136, 138, 148, 151, 156, 157, 161, 170–73
  oil in, 3, 6, 8, 9, 13, 14, 17–19, 46, 52, 54, 129, 131, 145, 161, 165, 176
  other resources of, 54, 156, 157, 159

politics in, 54, 56–59, 89, 99, 124, 137, 172

public opinion and, 15, 21, 24, 31, 46, 47, 50, 51, 54, 58, 75, 79, 83, 86, 91, 92, 105, 109, 112, 121, 122, 136, 137, 140, 156, 161, 171, 172, 174, 177, 178

wildlife in, 132, 148

Alaska Federation of Natives, 71, 78–80, 82, 85–90, 94, 100, 103, 106, 112, 120, 121, 168, 172

Alaska Native Brotherhood, 73

Alaska Review of Business and Economic Conditions, 74, 76, 99

Aleuts, 21, 44, 61, 63, 67, 71, 100, 117, 125, 128; *see also* Natives

Anaktuvuk Pass, 67–69, 75, 88, 94, 161

Anchorage, 23, 27, 31, 54–57, 59, 79, 91, 101, 139, 151
  described, 43–49, 51

Anchorage Daily News, 49, 50, 94

Anchorage Daily Times, 23, 91, 173, 174

Anderson, Clinton P., 111

Anderson, Robert O., 8, 11, 18, 122, 135, 142, 175, 176

Andrews, Rupert, 148

Arctic National Wildlife Refuge, 8, 132, 159

181

Arctic Slope Native Association, 32, 77, 78, 95, 103, 108, 169

Aspinall, Wayne, 70

Athabascans, 63, 72, 76, 78, 88, 90, 95, 99, 106; *see also* Natives; Indians

Atlantic Refining Company, 6, 10

Atlantic Richfield (Arco), 3, 6, 8–14, 17–19, 21, 23, 25, 27–35, 38–40, 59, 127, 130, 135, 138, 142, 155, 157, 173

Atwood, Robert W., 23–25, 33, 34, 38, 40, 55, 157

Barges, 29, 33, 156

Barrow, 5, 39, 58, 64, 75, 103, 105, 170

Bartlett, E. L., 75, 79, 81

Bellmon, Harry, 111

Bender, Jane, 94

Blair, John, 19, 20

Borbridge, John, 106

Brennan, Tom, 23–25, 27–29, 33–36, 38, 39

Bristol Bay, 160

British Petroleum (BP), 14, 17, 26, 27, 29, 31, 33, 39, 127

Brooks Mountains, 3, 27, 52, 67, 77, 94, 129, 141

Brower, David, 142

Bureau of Indian Affairs, 68, 72, 76, 88, 109, 117, 121, 169, 170

Bureau of Land Management, 74, 94, 145

Canada, 38, 64, 65, 132, 159

Caribou, 63, 66–68, 74, 103, 104, 132–34, 142, 145, 161, 162

Carter, Luther, 93, 141

Church, Frank, 111

Clark, Ramsey, 69, 70, 82, 99, 102, 106, 109–11, 113, 125, 168, 172

Cook Inlet, 46, 63, 131

Ecology, 21, 23, 131, 132, 135, 136, 140, 141, 143, 146, 147, 149, 155, 161, 162, 168, 172

Edmondson, Ed, 91, 106, 111

Environmental Defense Fund, 129, 158

Environmental Protection, 21, 23, 24, 31–34, 127–48, 152, 155–63, 168, 170, 172, 178, 179

Eskimos, 5, 21, 32, 38, 39, 61, 62, 64, 65, 67, 68, 75–77, 88, 94, 95, 100–106, 117, 123, 125, 128, 161, 169, 170, 177

Evans, Brock, 172

Fairbanks, 11, 24, 27–29, 62, 67, 69, 94, 95, 98, 101, 156, 174 described, 49–51

Fannin, Paul, 111, 112

Federal Field Committee for Development Planning in Alaska, 59, 63, 64, 66

Fitzgerald, Joseph, 59, 81, 87, 89, 113, 142

Friends of the Earth, 129, 142, 158

Goldberg, Arthur, 82, 91

Gravel, Mike, 57, 170

Gruening, Ernest, 59, 79

Guion, Charlie, 25–27, 36, 41

Gulf Oil Company, 25, 26

Hart, James, 99, 175

Hatfield, Mark, 111

Hensley, Willie, 78, 90, 101

Hickel, Walter, 4, 31, 48, 58, 59, 79, 80, 82, 92, 93, 98, 108, 111, 128, 129, 135, 136, 139, 145, 149–51, 156, 158, 159, 170–73

Hickel Highway, 156, 157

Hippler, Arthur, 47, 124

Hobson, Eben, 78, 103

Hopson, Alfred, 5, 64, 105

Houff, William, 141

Hughes, George, 140, 163, 164

Humble Oil Company, 3, 6, 11, 14, 15, 25, 27, 29, 127, 175

Indian Claims Commission, 109, 110

Indians, 21, 48, 61–63, 67, 69, 76, 100, 108, 109, 119, 123, 125, 126, 128, 170, 171, 178; *see also* Athabascans; Tlingit-Haidas

Institute of Social, Economic, and Government Research, University of Alaska, 74, 124

Interior Committee, U.S. House of Representatives, 11, 70, 80, 101, 102, 106, 111, 128, 150

Interior Committee, U.S. Senate, 58, 69, 80–82, 101, 109, 115–18, 120, 128, 137, 138, 140, 146, 150, 152, 155, 162, 163, 166, 167, 169, 179

Interior Department, U.S., 72, 76, 80, 92, 95, 98, 108, 111, 127, 128, 131, 150, 151, 158, 159, 162–67, 170–73

Jackson, Henry M., 55, 81, 82, 87, 89, 106, 108, 110, 111, 113, 115, 118–20, 128, 138, 150, 151, 154, 162–65, 167, 170, 172

Japan, 128, 151, 166

Jessen's Weekly, 94

Johnson, Lyndon B., 80, 108, 134

Juneau, 90, 102, 172, 173
    described, 56, 57

Kelly, Thomas, 58, 138

Kenai Peninsula, 10, 46, 54–57, 130

Ketzler, Alfred, 89, 96, 99, 107, 108, 113

Lachenbruch Report, 149, 150, 152–54

Land freeze, 79, 80, 83, 98, 99, 113, 128, 162

Lease sale of September *10, 1969,* 11, 21, 27, 51, 81, 90, 113, 125, 173, 175

Levy, Walter J., 15

Meeds, Lloyd, 11

Middle East, 15, 16, 18, 19, 166

Miller, Keith, 11, 89, 93, 138, 151, 156, 157, 171–73

Mills, Wilbur, 130

Milton, John P., 129, 130

Minto, 76, 78, 96

Mobile Oil Company, 57, 58, 105

Morel, Buddy, 139, 143, 145, 147, 153, 154

Murkowski, Frank, 123, 137, 138

Native land claims, 11, 21, 61–126, 169, 177–79; *see also* Aboriginal use and occupancy

Natives, 21, 45, 50, 61, 62, 66, 67, 69–73, 77–83, 85–90, 94–96, 98–102, 105–14, 118–26, 128, 132, 169, 177–79; *see also* Aleuts; Athabascans; Eskimos; Indians; Tlingit-Haidas

Naval Petroleum Reserve Number Four, 5–8, 129

Nelson, Gaylord, 166

Ninth Circuit Court of Appeals, 79, 99

Nixon, Richard M., 79, 134–36, 145, 151, 168, 172

Nixon Administration, 92, 127, 129, 135–37

North Slope (Arctic Slope), 157, 159
    background of oil strike on, 3–21
    described, 23–41
    land on, 129, 130, 141
    natives on, 65, 77, 78, 105, 119, 120, 169
    oil of, 43, 45, 46, 48, 49, 52–55, 57, 81, 93, 94, 99, 105, 127, 128, 155, 160, 165, 166, 168, 170, 172, 173, 175, 176, 178, 179
    wildlife of, 133

Notti, Emil, 71, 78, 85, 87, 89, 91, 92, 95, 112, 120–24

Oil import policy, 18–20

Oil industry, 7–10, 12–14, 16, 17, 19–21, 25, 26, 31, 46, 47, 53,

Oil industry (*cont.*):
54, 56–59, 83, 91, 94–96, 109,
122, 123, 129, 133–37, 139,
141–43, 147–49, 152, 159, 161,
163, 166, 167, 172, 174–79
Oil spills, 131, 134, 135, 145, 160,
163–65

Paneak, Simon, 68, 69, 75
Paul Frederick, 32, 96–97, 108
Paul, William, 73, 77, 78, 95
Pecora, William T., 8, 152, 172,
173
Permafrost, 26, 32, 34, 35, 40, 41,
133, 139, 149–55, 157, 170
Profits on North Slope oil, 155,
168, 171, 173, 176, 178, 179
amounts, 16, 17, 35
Prudhoe Bay, 3, 5, 6, 11, 17, 21, 32,
33, 76, 93, 104, 129, 130, 140,
159, 174
Puget Sound, 14, 18, 172

Ream, Louis, 19, 20, 142, 151, 155
Richfield Oil Company, 6, 10, 55,
56
Rock, Howard, 70
Royalties, 11, 21, 53, 56, 82, 83, 88,
90, 111–13, 115, 117–19, 173
Russia, 6, 43, 44, 62, 63, 106, 107

Sage, Bryan, 141
Salmon canning industry, 54, 59
Santa Barbara, 131, 135, 136, 160
Seattle, 19, 20, 43, 77, 141, 155,
172, 174
Sierra Club, 142, 146, 149, 172

Standard Oil Company (New Jersey), 17, 19, 173
Stevens, Ted, 57, 91, 105, 115, 117,
137, 170

Tacoma, 120
Tanacross, 72–76
Tankers, 14, 17, 18, 20, 131, 160
Thomas, James, 71, 121
Tlingit-Haidas, 63, 71, 73, 74, 106,
107; *see also* Indians; Natives
Train, Russell, 128, 129, 136, 145–
47, 162, 166, 167
Trans-Alaska Pipeline, 13, 14, 21,
48, 57, 94–98, 127–68, 169–
75, 178, 179
Trans-Alaska Pipeline System
(TAPS), 14, 95–99, 127–29,
138–43, 145–48, 150–57, 161,
167, 170, 171, 173, 179
Tundra Times, 71, 72

Udall, Stewart, 79, 80, 128, 131
U.S.S. *Manhattan,* 14, 58

Valdez, 13, 14, 20, 48, 127, 133,
151, 160
Van Ness, William, 153, 157

Washington, State of, 18, 172
Weeden, Robert, 146–48, 155, 157,
160, 161
Weissberg, Barry, 132
Wilderness, 23, 24, 31, 51, 127, 129,
137, 138, 156, 158, 161, 162,
168, 178, 179
Wilderness Society, 129, 158